TOXIC

Neil Nathan, MD

Victory Belt Publishing Inc.
Las Vegas

First Published in 2018 by Victory Belt Publishing Inc.

ISBN-13: 978-1-628603-11-8

Cover Design by Justin-Aaron Velasco
Interior Design by Charisse Reyes and Elita San Juan

Printed in Canada
TC 0520

Contents

Dedication

It is said that the eyes are the window into the soul. I can't speak for others' eyes, but I can tell you that Cheryl's are. Her blue-grey eyes, with a ring of brown, sometimes green in certain lights, were the first thing I saw when I looked at her...and I've been enchanted ever since. They are special eyes...they are luminous, full of light and love...but they are not challenging, hard eyes. They don't ask or demand anything; they radiate her unconditional love. While she has long maintained that she is "garden variety," she is anything but. What she is, is present. She is the essence of just "being." It is a rare individual who can manifest just being, and she does. She brings that into her every moment, and it is special beyond imagining. She creates the safe haven of our home and our lives in which I can create and work at healing. That space allows me to keep the fires burning and return to her each evening for the peace and serenity of our home. She is my muse, and without her, this book would not be possible.

I also dedicate this book to my beautiful grandchildren, Avi, Anjali, and WiLo. You are our future, and this book is written, in part, so that you and your generation may find in these pages a blueprint for healing.

And for the thousands of patients who have placed their trust in me, I am eternally grateful. Knowing full well how little I (we) know about these controversial and complex illnesses, you mustered your belief in my tenacious quest to find you answers and allowed me all sorts of latitude to do so. I hope and pray that I have deserved your faith in the service of healing. For most of you, I believe we have succeeded; for the others, know that I will keep searching.

Foreword

We are rapidly entering a new chapter in the history of medicine. In the old medicine, most diseases were caused by acute problems stemming from infections, acute exposure to toxins, or physical injuries. Today, healthcare workers spend most of their time caring for patients with chronic illnesses that last for years. These new chronic illnesses have both genetic and environmental causes. They are "ecogenetic" diseases that occur because the chemistry of the environment is changing faster than our genes can adapt. Eighty to ninety percent of clinic visits today are follow-up visits for patients with illnesses that modern medicine can make a little better with conventional treatment, but with rare exceptions cannot cure. To prevent these disorders, we need to clean up the environment. But to treat these disorders, we need a new kind of medicine.

Dr. Neil Nathan has woven into the pages of this book the first descriptions of a whole new book of medicine. While the *First Book of Medicine* was focused on treatment for acute illnesses, the *Second Book of Medicine* will collect the science and art of caring for patients with chronic illnesses. Using examples drawn from a lifetime of practice in the care of patients with Lyme disease, mold toxicity, mast cell activation, *Bartonella,* porphyria-like crises, and many other complex clinical disorders, Dr. Nathan guides the reader along in this complex journey to better health care for patients who all too often have fallen through the cracks.

Highly sensitive patients often face denial and dismissal from mainstream medicine. This occurs in part because thirty years ago, when today's physicians were in medical school, highly sensitive patients were so rare as to be relegated to "anecdote" and were unable to be studied systematically. The majority of physicians practicing today were never taught about these once rare but now common disorders. Dr. Nathan begins to right this wrong with this book.

Breakthroughs come when two requirements are fulfilled: 1) the repeated application of the old ways has failed, and 2) someone has the courage to try something new and then follow the data. For example, when a standard blood test like a comprehensive metabolic panel (CMP) is "normal" in a child with autism spectrum disorder or in an adult with myalgic encephalomyelitis/chronic fatigue syndrome (ME/CFS), it is not helpful in guiding either diagnosis or treatment. Repeating the CMP three times does not give a doctor three times more information. The patient is clearly suffering, but the old tests are blind to the problem. New tests are needed. Like a new lens, advances in medical technology are giving doctors a new way to "see" and diagnose disease. Once we can "see" a problem, we can begin to understand it.

Hippocrates once said, "The art is long, . . . and decision difficult." We are fortunate to have Dr. Nathan as our guide. His wisdom, clinical acumen, kindness, compassion, careful listening to his patients, and lifetime of experience have sparked medical insights that have the power to inspire a new generation of physicians. In the pages of this book, we encounter for the first time a new understanding of the pathophysiology of chronic illness built upon a deep understanding of the cell danger response. New treatment approaches are also described, creating a practical toolkit for managing complex, highly sensitive patients with chronic illness. Some of the tools described today will change and be expanded over time, but these first steps will help guide physicians and patients alike to a more hopeful future, where chronic illness can become a thing of the past and recovery of health becomes the routine outcome of new medical care.

Robert K. Naviaux, MD, PhD

Professor of Genetics
Biochemical Genetics and Metabolism
Departments of Medicine, Pediatrics, and Pathology
Co-Director, The Mitochondrial and Metabolic Disease Center (MMDC)
UCSD School of Medicine

The Rebooting Paradigm for Really Sensitive and Toxic Patients— Helping Them Get Well

> **" I am the Lorax. I speak for the trees. "**
>
> —Dr. Seuss

It has become increasingly clear to me that one of the biggest obstacles patients who have complex medical illnesses face is that the traditional system of medical care may not apply to them. If you are not feeling well for any reason, of course you want to put your faith in your doctors and the hospitals they serve. Faith is an integral component of the healing process, and I would hope that your faith in a system of health care is well founded. After all, over many years, you and your family may have been helped enormously through issues such as childbirth, surgeries, infections, and health emergencies. Your doctors, whom you have come to like and admire, have helped you through all of these issues. So now that you (or a loved one) have inexplicably come down with some as-yet-unnamed condition, surely those same dedicated healthcare providers won't let you down.

But some of the illnesses we are seeing today seem to be of a different breed entirely. Patients present with symptoms that are more global, more diffuse, and harder to pin down. These symptoms arise in so many different areas of the body that they don't seem to make sense. Your family physician has run as many tests as he or she can think of, and nothing is showing up. He or she may have referred you to rheumatologists, infectious disease specialists, gastroenterologists, orthopedists, neurologists, and pain specialists and finally, in desperation, has sent you to a psychiatrist for evaluation. Everyone shakes their head, baffled.

To make matters worse, because the root of the problem is so unclear, the growing implication is that somehow the illness is not "real"—it is all in your head. As doctors communicate that message to you and your family, your family begins to assume the same thing—that you are somehow under too great a stress or have some perverse "need" to be ill (which is known as *secondary gain*) or are malingering. Deep inside, you know that this is not true—it is *not* all in your head. You are sick, but you don't know why, and neither, it seems, does anybody else. Where else can you turn for help?

Sooner or later (but please, for your sake, I hope it is sooner), you will begin to wonder whether the system of medicine you have been using is up to the task of diagnosing and treating you. It seems kind of obvious that if you took a malfunctioning car to the same mechanic and he or she kept working on it without repairing the problem, at some point you would take the vehicle to another mechanic for another opinion. But doctors are not often viewed in the same way. It is very difficult to wrap our heads around the possibility that these bright, well-trained, caring people are not figuring things out, and neither are the colleagues with whom they consult. So you begin to believe that perhaps it *is* all in your head, or maybe you have a condition so rare and so terrible that no one, anywhere, will be able to diagnose and treat it. You are doomed to get worse and worse and dwindle away into a world of endless suffering and debilitation.

I wrote this book to show you that there is indeed another approach. If you are reading these words, you have already realized or have begun to realize that what you have been doing is not working and that you need to look for other information, other diagnoses, and other treatments.

In my previous two books, *On Hope and Healing* and *Healing Is Possible,* I outline a basic approach to the diagnosis and treatment of patients with what we are now calling *complex medical illnesses*. Examples include, but are not limited to, fibromyalgia, myalgic encephalopathy/chronic fatigue syndrome, Lyme disease with its co-infections, autism spectrum disorder, neurodegenerative diseases (such as multiple sclerosis, ALS, and Parkinson's disease), mold toxicity, heavy metal toxicity, chronic viral infections, PANS (Pediatric Acute-onset Neuropsychiatric Syndrome), PANDAS

(Pediatric Autoimmune Neuropsychiatric Disorders Associated with Streptococcal infections), and many others.

For reasons I cannot pretend to understand, there is a considerable rift in medical information such that most of these named illnesses are quite controversial. This rift leaves patients in limbo between warring philosophies. I view this as unfortunate; patients should never be forced to choose sides.

On one side of this rift are those medical authorities who doubt the existence of these named conditions or feel that there is "not enough good science" to evaluate and treat them. Therefore, they believe it would be irresponsible to make any of these diagnoses and would be uncomfortable providing "unproven" treatments. On the other side are those who, like myself, believe that making these diagnoses is just the first step in helping patients get well, because they suggest a wide array of possible therapeutic approaches. Having seen these treatments work well for the vast majority of my patients over a twenty-five-year period, I believe we can offer legitimate hope for recovery for those who have been left by the wayside.

Thousands of patients have been restored to health by these newer concepts and treatments. However, we are beginning to see yet another incarnation of these disease processes. The patients who come to see me today are a quantum leap (or two) sicker than the ones who came to me in the late 1980s. My current patients have been sicker for longer periods before receiving a diagnosis or treatment. It is common for patients with Lyme disease to have been ill for *five to twenty years* before someone comes up with the diagnosis. Some of my patients have been treated with the best protocols currently available but are slowly getting worse—more fatigued, more brain-fogged, more crippled by pain, more anxious, more depressed, and more sensitive to everything: light, sound, chemicals, medications, herbs, supplements, and even tiny doses of homeopathic remedies or electromagnetic frequencies. This increased sensitivity is even more baffling than the symptoms that precede it. Physicians who are not familiar with this process shake their heads and communicate to these unfortunate patients that it is "impossible" to react badly to whatever treatment or dosage has been prescribed. They talk to the patients' families and make it clear that this kind of sensitivity simply does not exist and (here we go again) is clearly "in the patients' heads."

I understand this thinking. When I first saw my first patient who reacted intensely to minuscule doses of hormones that I knew she needed (based on both laboratory testing and her symptoms), I did not get it, either. While I hope I did not let my skepticism show, I admit that deep down I questioned the veracity of her experience. But this was just the tip of the iceberg. Over time, as I continued to work with patients whom no one else knew what to do with, I saw more and more of these ultrasensitive individuals. Most important, I began to understand that these reactions are real, not psychological, and must be taken seriously if these patients are to be helped. The concept that any medication, herb, supplement, homeopathic remedy, or treatment of any kind is so safe and gentle that no one could possibly react to it is simply wrong. My patients have reacted badly to virtually everything you can think of, even probiotics. Because these unusually reactive and sensitive patients now constitute 70 percent or more of my practice, I can report with certainty that each and every substance given to one of these individuals must be provided cautiously, because unique responses are the norm, not the exception.

Now that I understand this and have had years of experience in working with hundreds of such individuals, I feel the need to share that information with both my colleagues and the general population. For those medical practitioners who think that these patients are rare

and that they will never encounter them in their practices, please, *please,* think again. As our world becomes increasingly toxic due to the prevalence of chemicals, heavy metals, genetically modified foods, radioactivity, and electromagnetic radiation, we are beginning to realize that this is a growing epidemic.

Sensitive patients, having already grasped that if they share their symptoms with their healthcare providers, they are likely to be labeled as psychologically unstable, are understandably reluctant to do so. The single most important aspect of approaching these patients is *to believe them.* They are not making up their symptoms. They are suffering. They might look okay on the outside, but inside they feel awful. Even a family member or friend cannot assume that if a person looks fine, he or she *is* fine. In fact, most of my patients, in an effort not to engage the skepticism of others, make great efforts to look as normal as they possibly can. Paradoxically, this makes it more difficult for others to take their complaints seriously. These patients are between the proverbial rock and a hard place: if they allow their suffering to show, they are labeled as "whiners," and if they don't, no one believes them.

I *do* believe them. No one would make all of that up. My job as a physician is to elucidate the causative factors that are making them so sensitive and to treat those factors in the proper order. This is not an easy task, and it is not for the faint of heart or for those without patience. Central to this effort is making the correct diagnoses. Note the plural form of *diagnosis.* The older medical model of finding the singular cause of a particular illness is passé. These illnesses are caused by years of biochemical and medical dominoes—a patient experiences an illness, perhaps a virus, or perhaps a motor vehicle accident, or perhaps simply childbirth, and for whatever reason is not fully restored to health. In a mildly compromised state, he or she is more susceptible to another illness, then another, and then another, until the patient reaches my doorstep. We need to work through these steps to clarify all the components, determine which of the inciting factors is the main one that is interfering with the healing process, and slowly and gently begin to reverse it.

That is what this book is about. My intention is to share with everyone who wants to listen how to make these diagnoses, tease them apart, and then begin the treatment process. The patients who have come this far down the road need a more profound form of treatment than the ones used successfully with other patients; they need to be "rebooted." In these pages, I will make every effort to make this rebooting process clear, because, once understood, it will lead to a host of integrative, alternative, and even conventional treatments that can help accomplish this goal. It will require, on everyone's part, a lot of time and patience, slow and gentle efforts, and, yes, a financial commitment as well.

Because I have seen this process work for hundreds of patients, this is also a book of hope—realistic hope for really, really sensitive and sick patients.

Let's get started.

Disclaimers

First, let me emphasize that I have no financial ties to any company, individual, product, supplement, test, laboratory, or website mentioned in this book. If I refer to a specific product or test, it means that I have used that product or test with a great many patients and have come to appreciate how well it works. I hope that makes all such recommendations as clean as I can make them.

As excited as I hope you will be to try the treatments I describe, please understand that all of them, under some circumstances, present some risk or danger. *Please do not embark on any treatment without the assistance of someone who is medically trained, knowledgeable, and experienced in the details of the testing and treatments described in this book.*

This book is not intended to be a treatment manual, but rather a starting point for understanding your illness and a catalyst for initiating progress in your medical journey.

Throughout this book, with a few exceptions that are clearly noted, I have changed the names of patients, all of whom have consented to have their case histories presented herein. I have done so to protect their privacy, but I have taken care to present their medical information as accurately as possible.

As medical information evolves, so does the work in progress that represents our life stories. By the time this book has gone to print, some of the details of this medical information may have changed, along with some of the stories that my patients have related. What you read in this volume are the best, most current truths to which I have access at this time, subject to change at a moment's notice. I will do my utmost to stay on the cutting edge of these truths.

PART 1

OVERVIEW

These first two chapters are intended simply to introduce patients and physicians to the subject matter of this book.

Let's start with why anyone would want to read this book. It seems to me that this is as good a place to start as any. Before you spend a great deal of time studying a whole lot of information, I want you to be sure that this information applies to you. Later chapters delve into biochemistry, epidemiology, toxicology, anatomy, and physiology in more detail, and I hope to prepare you for this journey so that you will understand it and want to learn even more.

Given how little is known about the causes and treatment of complex medical illnesses, to even consider writing this book is a bit presumptuous on my part. But so many of the patients I treat are so lost, without a compass or rudder, that they have no idea how to even begin this journey. It is my intention to give you a solid starting point. I anticipate that in the near future, we will learn a great deal more about this subject, and the information in this book will be replaced by newer, better science. I hope to be around to be a part of that, but in the meantime, I also hope to provide a framework from which future physicians can evolve this field.

Very simply, Chapter 1 gives an overview of what will be discussed in this book. Chapter 2 starts to dig into toxicity and sensitivity and how we should begin to consider those processes in chronically ill patients who are not responding well to the usual therapeutic interventions.

CHAPTER 1

Why Would Anyone Want to Read This Book?

I specialize in diagnosing and treating patients who find that few practitioners know what to do with them. This field of medicine has no name as yet; I have been calling it *complex medical problem solving* for some time. I didn't start out in this field; in fact, this field did not really exist until recently.

I started out as a small-town family physician, delivering babies, working in the ER, and doing a little surgery after completing my work with the Indian Health Service in 1974. (If you want to do the math, since completing my internship at San Francisco General Hospital in 1971, I have been in practice for forty-seven years.) Early on in my career, for some inexplicable reason, I became interested in those patients who did not have a clear diagnosis or weren't improving with conventional approaches. My lifelong fascination with what are now called "outliers" led me to study a wide variety of medical and healing traditions, always looking for answers for these more complicated patients. I detail that journey in my first two books if you are interested.

However, since I wrote *On Hope and Healing* in 2007, there has been an unfortunate epidemic that has rendered patients even sicker than the ones I describe in that book. Not only are they more compromised in every way—they have even more fatigue, pain, cognitive dysfunction, anxiety, depression, and unusual and hard-to-describe symptoms—but these patients have become, over time, exquisitely sensitive to their environments and simultaneously to the medications, supplements, and other treatments that helped so many of my patients in the past. This hypersensitivity has made treatment, which is already complicated and difficult, even more challenging. It has necessitated that we find new ways to understand and contend with this emerging wave of compromised patients.

We have discovered that, for most of these patients, toxins (most commonly mold) and infections (most commonly Lyme disease with its co-infections) are the major causes. I will discuss these causes in detail in Part 2.

How many people are we talking about here? You might think that what I am describing does not apply to too many folks, but in August 2013, the U.S. Centers for Disease Control announced that there is indeed an epidemic of Lyme disease in this country: to wit, more than 300,000 new cases a year! To put this number in perspective, about 40,000 new cases of HIV are discovered annually. In addition to that large group of individuals, the prevalence of multiple chemical sensitivities (MCS) is noted in recent research to be 3.2 percent of the population, which may include as many as ten million people. So, not rare at all! Why might you not know about it yet? Because the unfortunate individuals who are suffering from these symptoms have learned the hard way not to tell too many of their family members or friends about their symptoms because of the response they get:

Now, let me get this straight, honey. You are tired all the time, can't read or see well, and can't focus or concentrate or think sometimes. You have a headache, joint and muscle pain, anxiety and depression, ringing in your ears, odd electrical and vibration sensations going up and down your spine, nausea and diarrhea, and numbness and tingling on your chin, and you are sensitive to just about anything your doctor wants to give you. Do I have that right, dear? But you look okay to me!

The implication from family, friends, and, worst of all, physicians and other healthcare providers is: *Really?* Translation: It's all in your head. So please do not be surprised if someone who suffers from all this is reluctant to talk about it. From years of experience, he or she has learned that the conversation rarely goes well.

To understand my perspective, it is important for you to realize that I have an unusual medical practice. My patient examples, my information, and my bias come from working with this uniquely ill group of individuals. While I believe that this information is of some general value, please keep in mind that I am writing primarily for these unusually sensitive and toxic patients and their healthcare providers. I hope that you will not take this information out of context unless you, or a loved one or friend, fit this model. On the other hand, if physicians and consumers understand how people who are really sensitive should be treated, then treating those patients with stronger constitutions will be far less difficult.

Let me present a fairly typical story of one of my patients so that you can begin to grasp both the complexity of these medical illnesses and these patients' responses to treatment. This first story is rather long, but it illustrates so many of the important components of diagnosis and treatment that I hope it resonates with you.

KAREN'S STORY

Karen first came to see me in October 2010, at the age of forty-seven. She had been essentially healthy until January 2008, when she returned home from a hike in northern California with her dogs and discovered that her pets were covered in small seed ticks. She does not recall receiving a tick bite herself, but two to three weeks later, she came down with a flulike illness involving intense neck, foot, and spinal pain and fatigue. She was fortunate because her family physician realized that the issue could be Lyme disease (an unusually quick diagnosis), but when placed on antibiotics, Karen experienced severe gastrointestinal symptoms and got progressively worse. By August, she had become weaker and more fatigued and required a walker to move about. Within a few more months, she had developed an extreme sensitivity to scents; mere exposure to perfumes or certain chemicals would cause immediate collapse and a brief loss of consciousness. Her doctor continued to treat her for Lyme disease, which included the use of a Rife machine (an electrical device that delivers specific electromagnetic frequencies to a patient with the intention of treating specific infections; see page 90) and intravenous fluids, which did help. However, by March 2009, Karen required the use of a wheelchair, her sensitivities (multiple chemical sensitivities, or MCS) had escalated to include exposure to fluorescent light, and she had developed intense anxiety with panic attacks and severe depression.

A complete neurological evaluation that included an MRI was negative for multiple sclerosis or any other clear cause of her debilitating symptoms. She continued to be treated by two Lyme Literate physicians (LLMDs) with antibiotics aimed primarily at Lyme disease (caused by Borrelia *bacteria) and babesiosis, a common co-infection of Lyme disease caused by a parasite and similar to malaria. Despite receiving expert care, her condition continued to deteriorate, and by May 2010, Karen began exhibiting spasmodic twitching and jerking motions (called* dyskinesias*) and had lost the use of her right hand. She was unable to read or walk.*

It was shortly thereafter that she came into my office for the first time. The overriding question during that first visit was, as always, "What are we missing?" Good testing from the IGeneX laboratory, the Western blot, showed clear Lyme disease and the presence of Babesia. However, Karen's symptoms of chemical sensitivity, anxiety and depression, and unusual neurological activity in the form of dystonias and dyskinesias, with intense fatigue and cognitive impairment, pointed to other possibilities. That collection of symptoms is not particularly characteristic of Lyme disease or babesiosis but is much more common in another Lyme co-infection, bartonellosis, and also is found in the presentation of mold toxicity.

To test this hypothesis—that the primary issue Karen was facing was likely Bartonella and/or mold toxicity—we did a confirmatory Bartonella challenge test using the Byron White tincture A-Bart. After receiving only three drops of this tincture, which is relatively specific to Bartonella, Karen had an intense Herxheimer reaction, with an intensification of her fatigue, cognitive impairment, and anxiety. Recognizing that even minuscule doses of any therapeutic material could make her worse, we started her on one drop of A-Bart once a day and slowly increased to one drop twice a day, which was as much as she could comfortably take. We added a daily HH supplement (a form of Houttuynia) and a small amount of Argentyn 23, or hydrosol silver. As she very slowly stabilized, we added the first antibiotic specific for Bartonella, the sulfa drug Bactrim. A typical dosage of Bactrim would be the double strength (DS) dosage twice daily, but for Karen, we started with a half dose once daily in August; it was November before she could take it twice a day. She improved and got stronger, and a month later we were able to give her the full dose of Bactrim DS twice daily. At about this time, we also began LENS treatments (see Chapter 9). Karen was no longer in a wheelchair but still could not leave home without the risk of chemical exposures that would cause her to crumple and have spasmodic dystonic reactions in church or at the grocery store.

Treatment for Bartonella typically requires the use of more than one antibiotic. After Bactrim, the best next antibiotic is usually rifampin (Rifadin), but even tiny doses set Karen back. So we turned to a low dose (250 milligrams) of levofloxacin (Levaquin), which fortunately she could tolerate, and she continued to make slow but steady progress into June 2011. We tried adding Babesia treatments, but none of those, even in minuscule doses, were tolerated. We began LDA treatments (see page 268) to quiet her immune reactivity in July 2011, along with the homeopathic supplements PSY-stabil, which helps quiet the nervous system's reactivity, and SyAllgen, which helps settle down the immune system's reactivity (but is no longer available).

By August, Karen could tolerate 150 milligrams a day of rifampin (the usual dosage is 300 milligrams twice daily), and we were able to increase this dosage very gradually. We began to explore the possibility that she also had mold toxicity, and a visual contrast sensitivity test was clearly abnormal. (See page 59 for more on this test; Karen's results are shown in Figure 1.1 on page 22.) It was not until December that we were able to get her to test her home with mold plates, and she realized that there was a significant mold infestation.

By now, Karen had slowly worked up to nine drops of A-Bart twice daily; her LENS treatments were enabling her to get out of the house occasionally and begin the process of rejoining her community. Given her extreme sensitivity to all treatment efforts, we began to treat the mold toxicity with intravenous phosphatidylcholine. At first she could tolerate only 1 cc (the usual dosage is 5 to 15 cc), but week by week, the dosage was increased to 5 and then 10 cc.

By March 2012, Karen could finally take a half capsule of colesevelam (Welchol), an important binder for mold toxins, and the mold in her home had been remediated. By September 2012 (notice how long it took to increase some of these dosages!), she could tolerate a whole capsule of Welchol, thirteen drops of A-Bart twice daily, and 150 milligrams of rifampin twice daily, and she continued with LDA treatments.

By December, she could handle two capsules of Welchol a day and 300 milligrams of rifampin twice a day, and she was able to begin the methylation protocol that previously had been off-limits, as the tiniest amounts of vitamin B12 and methylfolate set off her nervous system. (See Chapter 14 for more on methylation.)

By June 2013, Karen could drive herself to the grocery store and to church and was able to bicycle again with only a minor risk of dystonic reactions. She began meditation and the use of affirmations (see Chapter 20), which, interestingly, set off another round of neurological spasms. As she was getting stronger, she was also becoming more aware of earlier psychological traumas that she was now, with the help of a therapist, able to process and release. She began to make more rapid improvements; she was able to resume part-time work and stopped taking all antibiotics. By December, she had only a mild blurring of vision and minor abdominal symptoms. By February 2014, she began Annie Hopper's Dynamic Neural Retraining System (see page 156) and was able to attend and enjoy her daughter's wedding and to fly to another city and stay in a hotel overnight to take courses for several days, which would have been unthinkable previously.

I continue to follow Karen, but she is basically well now. She has completed 100-mile bicycle races, scuba dives in the waters off of San Diego, and is working full-time.

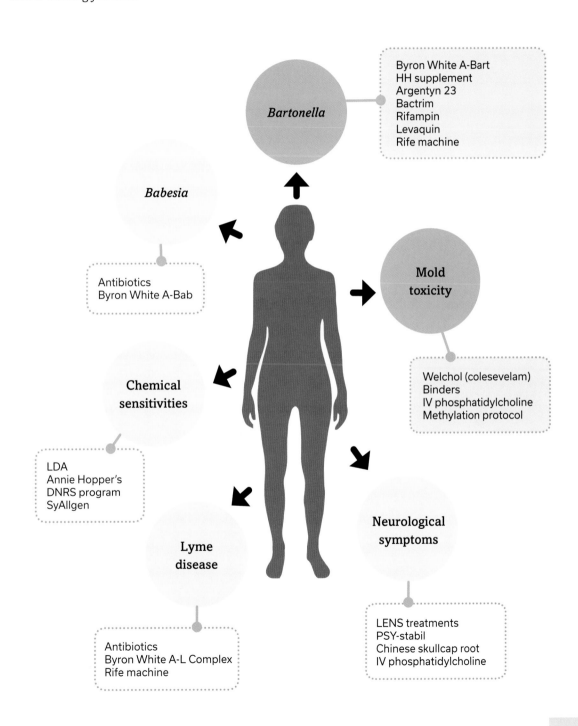

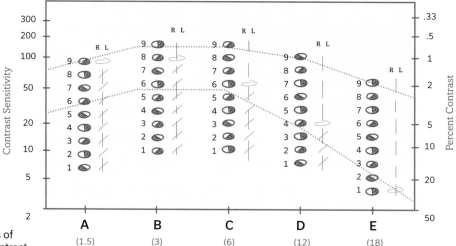

Figure 1.1. The results of Karen's initial visual contrast sensitivity test. To interpret this chart, first note the hand-drawn checks and ovals. A check means that black-and-white component could be seen clearly; an oval means it was not seen. Anything that does not even reach the lower line (an oval) means that the individual did not pass this test. For Karen, nothing was visible in row E, and she was barely able to see up to the line in row D, which means that she failed this test and likely had some form of biotoxicity.

Karen's story demonstrates many of the ideas and concepts that are central to this book.

First, although she had been correctly diagnosed with Lyme disease and *Babesia,* these were not the primary issues that her immune system was wrestling with; *Bartonella* and mold toxicity were the primary issues. Critical to understanding how to approach sensitive and toxic patients is the simple principle of making the correct diagnosis. I call this identifying what the patient's body declares to be Public Enemy #1. Unless you know with precision that what you are treating is *central* for this patient (even if the patient has other, correct diagnoses), you will not make much progress. For example, working on curing a viral component when the primary problem is mold toxicity will not move the patient toward health and could potentially distract the patient's body's healing systems from their primary objective. I will continue to emphasize this point throughout this book. There are few things of greater importance than nailing the diagnosis so that you can begin the right treatment.

Second, Karen's story beautifully illustrates that we must listen carefully to every patient so we know when we are going too fast. We won't make egregious mistakes by going slowly, but treatment that is too aggressive (such as too much medication too soon) can set a patient back light-years. The tiny, incremental doses of medications, herbs, homeopathics, and other treatments we used for Karen slowly but surely helped her to heal. Every time we pushed it too far, she got worse, and her body let us know that we needed to pull back, let her body settle down, and then proceed very gently. At its essence, caring for these kinds of patients is about listening to everything they say so that treatment is in sync with their needs and abilities to tolerate what they are given.

Please keep these principles in mind and remember that they are essential for those who are unusually sensitive or toxic, but might not apply to those who are just getting sick, are only mildly ill, or are constitutionally stronger. Healthier individuals often move along much faster and more easily and experience fewer setbacks. Treatments must be individualized to each unique patient.

To reiterate: This book is about how to approach, diagnose, and treat those patients who are unusually toxic or sensitive, or how to approach your care if you are one of these patients. These conditions are far more common than is currently appreciated by the medical establishment, who, by not believing the descriptions they are hearing, often abandon these patients to the psychiatric world. If we can understand how to treat these very complex patients, then treating others who are not as ill will be much less complicated.

Having treated hundreds of really, really sensitive patients over the past fifteen years, I can share how gratifying it has been to help people like Karen get their lives back. This experience has convinced me, beyond a shadow of a doubt, that the causes for this marked increase in sensitivity are not psychogenic but rather physical and treatable. So, if this description applies to you or a loved one, please know that there is hope for a complete recovery, regardless of how compromised you or your loved one has become.

CHAPTER 2

Sensitivity Versus Toxicity (or Both)

What's the Diff?

At the risk of offending the membrane physiologists and toxicologists who may read this chapter, I am going to attempt to simplify a very complicated subject. I do so in the spirit of giving you some ways to understand what we mean when I and many others throw around terms such as *toxicity* and *sensitivity*. I often see articles that discuss toxicity in a general sense, but I hope that you will begin to ask the important question: *"Toxic to what, exactly?"* I think this question is very useful from both a diagnostic and a therapeutic standpoint.

What Is Toxicity?

In her lectures to medical audiences, Patricia Kane, PhD, a pioneer in the understanding and treatment of cell membrane imbalances, projects slides showing electron microphotographs of toxic cells. You can see in these close-ups cell membranes that are chock-full of toxin. Some cells are so loaded that you can see molecules of toxin precariously attached to the outer membrane, ready to be dislodged with the slightest provocation. This visual can be very helpful for showing how easily the tiniest insult or stressor—a shower that is a little too hot or too cold, a barometric weather change, or exposure to a certain food, chemical, or scent—will knock loose those toxic molecules and send them into the bloodstream, exacerbating that person's symptoms.

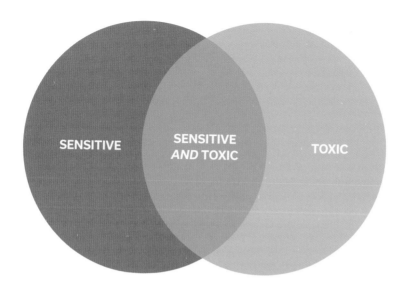

SENSITIVE

SENSITIVE
AND TOXIC

TOXIC

That the cell membranes are so built up with toxin is an obvious (but easily overlooked) problem: there is no room on the membrane for the membrane to be able to do much of anything else. It cannot get rid of additional toxins as they accumulate. It is, literally and figuratively, stuck. Once toxins have begun to saturate the cell membranes, these toxins, which can include mycotoxins (mold), heavy metals (for example, mercury or lead), and pesticides and herbicides (such as glyphosate), literally poison the body's systems of excretion and elimination, preventing them from doing their jobs.

To put it more directly, being poisoned prevents the body from doing what it needs to do—it simply cannot get rid of what is making it sick. This has profound implications for both diagnosis and treatment, as you will see in Chapter 3, which discusses mold toxicity.

To expand on this subject, some mold toxins are *ionophores*. An ionophore is a somewhat unique molecule that has a fat-soluble piece (which is technically called *lipophilic,* meaning "lipid-loving" or "fat-loving") on one end and a water-soluble piece (called *hydrophilic,* meaning "water-loving") on the other end, as shown in Figure 2.1. The importance of this designation is that these molecules can attach to, or blend in with, the fatty materials that constitute every membrane in the body or blend in with or dissolve in watery solutions.

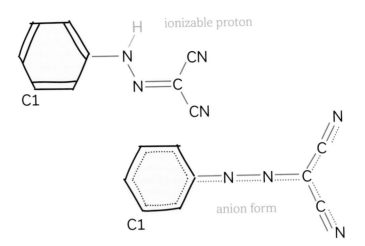

Figure 2.1. An ionophore has a lipophilic end and a hydrophilic end.

The body typically prevents various molecules from entering a cell by utilizing a complicated system of membrane physiology. A foreign molecule is not allowed to enter the cell unless it binds to specific receptors on the cell membrane. Ionophores, by their very nature, can bypass the cell's normal defenses and go anywhere they please—meaning that these toxins can enter any cell and travel through the bloodstream, and the body has very few defenses against them.

So, how *does* the body get rid of these molecules? In the case of mold toxins, if a person has the right genetic makeup (which exists in 75 percent of the human population), he or she can make antibodies to these toxins that will help bind the toxins so that the body can get rid of them. But for the 25 percent of the population that is not genetically engineered to make these antibodies, the only means to deal with these toxic molecules is to bring them to the liver, our major organ of detoxification, or perhaps to utilize other systems of elimination, including the skin, gastrointestinal tract, kidneys, lymphatic system, and lungs, which have the important job of finding a way to move the toxins out.

To reiterate, if these systems are literally poisoned, they may not be able to do their jobs, and toxins thereby can accumulate. This is the state of affairs for many of my sickest patients. Even if we can figure out what is making them toxic—be it a mold toxin or mercury, for example—we may have to address this shutdown of the organs of elimination before we can even begin to consider treatment. (See Chapter 15.) If we plunge into treatment with the intention of healing a patient *before the patient's ability to detoxify has been addressed*, there is a very good chance that the patient will only get worse.

Let Pierre's story help clarify what toxicity looks like.

PIERRE'S STORY

Pierre is a forty-four-year-old man who came to see me recently after becoming ill four years ago. He described increasing fatigue, insomnia, frontal headaches, nausea (occasionally with vomiting), brain fog, anxiety, and depression. Of particular concern was an intense sensation of tingling and numbness around his mouth (technically called perioral paresthesias*). Another physician had diagnosed him with Lyme disease, but antibiotics produced what Pierre called a "huge Herx." (A Herx reaction—short for Herxheimer, named after the physician who first described it—is a two- to three-day exacerbation of symptoms caused by the sudden release of toxins produced by killing a bacterium such as Lyme or* Bartonella*.) He became progressively worse. As he researched his condition, he discovered the possibility of mold toxicity. Upon reflection, he realized that a shower in his weekend home (which he visited regularly) had developed a leak, and he had been involved in the remediation process in which mold was uncovered and removed. Further reflection revealed other possible exposures to mold at work and in previous homes.*

Pierre persuaded his physician to do a RealTime urine mycotoxin test, which showed an ochratoxin level of 3.10 (normal is less than 1.8) and a gliotoxin level of 1.76 (normal is less than 0.5). (See Chapter 3 for a detailed discussion of these tests.) With this clear diagnosis of mold toxicity based both on symptoms and laboratory testing, we started treatment with the binders that are best for those two categories of mold toxins: bentonite clay, Saccharomyces boulardii, *N-acetyl cysteine (NAC), activated charcoal, and cholestyramine. (Binders are natural or pharmaceutical materials that have the ability to attach loosely to a specific mycotoxin and carry it out of the gastrointestinal system.)*

When we talked just one month after he began this treatment, Pierre reported marked general improvement, noting that "my strength is coming back." The perioral paresthesias that had been particularly bothersome to him were greatly reduced, and he was able to swim for the first time in a long time.

Here is a clear example of toxic exposure to mold. Over time, classic mold toxicity symptoms slowly amplified until specific treatment to rid Pierre's body of those toxins was provided. More rapidly than is typical, Pierre noted significant improvement that was directly connected to his treatment. I am happy to report that he continued to improve as we fine-tuned his treatment program. Simply put, toxin exposure had made Pierre ill. Treating that exposure began the process of helping him get well.

What Is Sensitivity?

Sensitivity refers to a nervous system that has become overexcited or hyperreactive in response to a wide variety of stimuli—sound, light, touch, foods, chemicals, odors, and even electromagnetic fields (EMF) for some people. A sensitive patient can have one or all of these manifestations.

You might be so sensitive to light that you need to wear sunglasses in most settings or spend most of the day in a darkened room. You might be unusually reactive to noises, jumping at unexpected sounds that others do not even respond to. You might develop food sensitivities that can include true allergies or mast cell activation. (*Mast cells* are immune cells that coordinate how the immune system and nervous system respond to infectious and toxic agents. Both mold toxicity and *Bartonella* infection sensitize these cells so that they become hyperreactive to stimuli that would ordinarily not be bothersome. An "activated" mast cell reacts intensely and somewhat randomly to a variety of stimuli with the release of large amounts of histamine and other chemicals that create an immediate onset of such symptoms as flushing, sweating, palpitations, abdominal pain, and diarrhea. See Chapters 5 and 12.) You might develop multiple chemical sensitivities (MCS) so that even the slightest exposure to certain odors (such as perfume, gasoline, and/or cigarette smoke) can, within seconds, cause profound fatigue, cognitive impairment, or more dramatic reactions, such as pseudoseizures and dyskinesias (spasmodic twitching and jerking motions that look like seizures, but without the accompanying EEG changes).

In Chapter 1, I told the story of Karen, who had some of these unusual neurological presentations. (See page 18.) I witnessed Karen walk down the hall of our clinic, doing fine, and then pass a staff member in the corridor and suddenly begin to twitch and writhe. Karen was unusually sensitive to Tide detergent, and the staff member had very recently washed her clothes with that kind of detergent. As unusual and extreme as this reaction may seem, I repeat that there is nothing psychosomatic about it. I urge physicians and patients' families alike to keep their minds open to these phenomena, and they will see that these events are predictable once the triggers are understood.

Some patients become unusually sensitive to EMF exposure and find that they can no longer work around computers. In rare cases, even Wi-Fi exposure, which is ubiquitous, can set off fatigue, cognitive dysfunction, and/or neurological events, making it difficult for individuals to find safe environments in which they can live. When the state of California mandated the use of smart meters, which often rely on Wi-Fi, my office saw several new cases of severe EMF reactivity in people who had been apparently healthy prior to the installation of those meters. I personally attended to several patients who had to leave the Bay Area and move to rural settings so that they could get away from all EMF exposure in order to function.

People who are not sensitive and therefore cannot fathom this level of sensitivity might read this and think that these reactions must be psychosomatic. *They are not.* This marked increase in the reactivity of the nervous system is a serious problem for many of my patients. It is hard enough for them to work through these difficulties without having to face skepticism, scorn, and derision from those who do not believe them. I have seen these dynamics tear families apart as patients' family members struggle to understand and take up positions that undermine communication and healing.

Sensitivity is created by a hyperreactive or hypersensitive nervous system. In a sense, you could think of it as a hypervigilant nervous system that is not convinced it is "safe." Given that sensitive individuals may experience debilitating neurological symptoms such as seizures at a moment's notice if exposed to a certain scent when passing someone else on the street, they understandably become worried about exposures they cannot control. More specifically, some of the parts of the nervous system that are involved in this hypervigilance are the limbic system of the brain (see the sections on Dynamic Neural Retraining and Polyvagal Theory in Chapter 9, beginning on page 156) and the ventral branch of the vagus nerve and associated cranial nerves.

Let Ruth's story clarify what sensitivity looks like.

RUTH'S STORY

Ruth was referred to me in 2010 by a colleague who had seen her for chronic fatigue syndrome, which he felt was caused by Lyme disease and possibly Bartonella *and* Babesia. *He also believed that there was a viral component to her illness, and she was being treated for all of those issues, which included the use of antibiotics. I discovered that Ruth also had some degree of adrenal fatigue (a deficiency of adrenal hormones created by prolonged stress) and incorporated that into her treatment program. Although she appeared to improve initially, over time it became clear that she continued to be extremely fatigued, with significant exercise intolerance (which we call* post-exertional malaise*), headaches, and cognitive impairment, which included brain fog and difficulties with focus, memory, and concentration.*

By 2014, seeing little improvement, I evaluated Ruth for mold toxicity, despite no clear evidence of current mold exposure ("but I grew up in a moldy basement," she later recalled). The test showed a clear elevation of trichothecene (over three times the upper limit of normal), so I began treating her for mold. This proved quite difficult, and Ruth became more and more sensitive to every treatment I prescribed. To be more specific, she could take only a few drops of the tinctures we began with and tiny amounts of nasal sprays. By April 2015, she told me, "I am reacting to pretty much everything." In addition to a worsening of her existing symptoms, she became depressed and physically weak. She reported an exacerbation of her condition when she took even one-sixth of one drop of cilantro (which is used for detoxification), and she noted an immediate onset of symptoms after eating a variety of foods, which suggested that she had developed mast cell activation syndrome (discussed in Chapter 5). Despite her best efforts, she could barely tolerate the treatments for mast cell activation and by early 2017 was very frustrated by her deterioration.

We then started Annie Hopper's Dynamic Neural Retraining System (see page 156), and Ruth's extremely reactive body finally began to quiet down. She responded beautifully to this form of treatment. Within several months, she was able not only to start tolerating the treatment for mast cell activation (which helped her greatly now) but also to resume specific treatment for mold toxicity. Over the course of that year, she got progressively stronger and better. She now reports a marked decrease in depression and sensitivity and significant improvement in energy and cognition. She is able to exercise for the first time in years.

While Ruth's story begins with toxicity, that toxicity eventually triggered severe sensitivities that profoundly impacted her treatment and her life. The DNRS program (see page 156), which deals primarily with sensitivity, was able to reverse this process and allowed Ruth to resume her healing journey. Her story also serves to introduce us to the combination of toxicity and sensitivity that is so common in my patients.

The Difficulty in Distinguishing Toxicity from Sensitivity

I have attempted to distinguish toxicity from sensitivity because they are not the same thing, even though clinically they sometimes *look* like the same thing. When a patient experiences an increase in toxicity, his or her symptoms can be identical to those in patients who experience reactions to stimuli to which they are sensitive. I think this has led many clinicians to lump toxicity and sensitivity together, frustrated by the difficulties of clarifying a complex situation. However, if we can separate them, we realize that they need to be treated differently, and this can be of overriding importance. Patients who are dealing with sensitivity, for example, are often unable to tolerate the treatment for toxicity. A sensitive patient will often react violently to the supplements and medications provided to remove toxins. Not recognizing the difference between toxicity and sensitivity, therefore, can inadvertently lead to making patients worse.

What makes this a little more complicated is that, as noted earlier, many patients are both toxic *and* sensitive. Worse, the presence of toxicity predisposes a person to sensitivity, and the increase in sensitivity predisposes that person to increasing reactivity to the toxins to which he or she is exposed. Therefore, it can be quite difficult to tease these phenomena apart with any degree of precision. Often physicians are forced to use a process of trial and error to figure out the right approach to each patient who presents with unique issues shaped by his or her own unique biochemistry and genetics.

The Epidemic of Increased Toxicity and Sensitivity

First, let me emphasize that the increase in toxicity and sensitivity that we are seeing is an unacknowledged epidemic. As I write these words, I do not feel that I am being overly dramatic. When I began treating chronically ill patients in the mid-1980s, I was unaware of these phenomena. By the late 1990s, I had seen my first few patients who were reacting to substances that I thought of as so benign that I questioned the validity of their reactions. Rare patients would complain that taking probiotics made them worse; others, that taking vitamin B12 made them worse; still others complained about tiny doses of homeopathic remedies. Having never seen this kind of reactivity before, I was baffled. These were not neurotic patients, but credible citizens from whom I did not expect "weird" reactions. I did not know what to do with these reactions, and unfortunately, I had little to offer either as explanation or in the way of treatment.

In a relatively short period—a year or two—these reports became more numerous, and I realized that I had to take them seriously. I still didn't understand what was happening to these patients, but I was past the point of questioning their descriptions. I was convinced that what they were telling me was accurate; I just didn't know what to do about it.

Empirically, using trial and error, I found that several types of treatments helped quiet down their hyperreactive systems. These included osteopathic cranial manipulation, frequency specific micro-current (FSM), homeopathic remedies, and low dose allergen therapy (LDA)—all of which are described in detail in later chapters. Over time, I began to see a pattern emerge: almost all of these patients had been exposed to mold or had Lyme disease. When we were able to successfully treat those conditions, the reactivity slowly dissipated, and the patients eventually recovered. In the ensuing years, several other components of sensitivity came to light, most notably mast cell activation and porphyria (described in Chapters 5 and 6, respectively).

Today, these patients make up the majority of my practice. I would estimate that 70 percent of my patients exhibit this kind of hyperreactivity. As we have slowly recognized the epidemic nature of Lyme disease and are beginning to realize the same for mold toxicity, it should come as no surprise that what was once rare is now not uncommon. If physicians are seeing these kinds of patients today (and not all do, of course; those who deny the existence of these illnesses and do not treat them may not see them at all), they will be seeing more and more patients presenting in this manner in the near future.

Why Now?

While there is no formal "scientific proof," a growing community of physicians agree with me that these phenomena are triggered by a significant increase in the toxicities of the world in which we live. We are exposed to thousands of chemicals that simply did not exist fifty years ago. We are also exposed to massive amounts of electromagnetic fields that did not exist fifty years ago; worse, we hold them in our hands, to our ears, and on our bodies all day long. We are continually exposed to all kinds of heavy metals, chemical pollutants, and radiation. (The effects of the Fukushima Daiichi nuclear disaster have not yet been remedied or studied.) Concerns about the effects of genetically modified organisms (GMOs) and food additives abound. We are, therefore, exposing our bodies to chemicals and radiation in amounts and varieties never before encountered in the history of humankind. Without adequate or complete knowledge, we are messing with our environment in ways that we are just beginning to understand.

I fear that unless we devote ourselves to fully understanding this and to limiting these exposures now, we will suffer damage that cannot be undone.

Canaries in the Coal Mines

Quite a few contemporary writers have alluded to what I am describing. It is not a new idea.

When coal miners explored a new vein of ore, they brought canaries down into the mine with them. Canaries are more sensitive to toxic gases than humans, so when the canaries keeled over in their cages without warning, it was a sign to get out of the mine immediately. I cannot help but view the epidemic increase in chronic fatigue syndrome, fibromyalgia, mold toxicity, multiple chemical sensitivities, cancer, and chronic infections (of which Lyme disease is an excellent example) as a manifestation of the toxicity of our modern world.

You can take the stance of "there but for the grace of God go I" and view these chronically ill individuals as unlucky. In a sense, they are. Their unique biochemistry and genetics predispose them to one or more of these illnesses. But take care, because they are just the tip of the iceberg. We are right behind them, as are our children and grandchildren, who will inherit this polluted world.

My good friend Stephen Buhner calls what I have just written a "mini-rant." I suppose it is. I appreciate your indulgence.

Not a Quick Fix

There is good news and bad news. The good news is that most of these long-suffering patients can be restored to health. The bad news is that doing so may take time.

The first question I am asked after I outline in my first visit with a patient what we will need to do is, "How long will this take?" My answer is, "I don't know." When I first started treating these kinds of individuals, it seemed reasonable to answer, "A year or more." While this was my standard answer, I found that patients heard only "a year" and the missed "or more" part. I emphasize that now.

At the 2018 meeting of the American Academy of Environmental Medicine, which was devoted to the treatment of mold toxicity, almost all of the speakers who have worked in this field for many years agreed that complete healing can take four to five years. Some patients respond quite quickly and get well within six months. A larger percentage of patients are well within one to two years. A smaller percentage do, indeed, take three to five years to get well. Many of those who take longer to improve have been unable to leave a moldy environment or have concurrent infections with Lyme or other microbes that complicate the healing process. On the other hand, many of the patients I see have been ill for fifteen or twenty years, and if it takes a few more years for them to get well, so be it. Patience and perseverance are essential components of the healing process.

Having written these words, I want to emphasize again that, at its heart, this is intended to be a book of hope. For those of you (or your patients or loved ones) who have entered this realm of sensitivity and toxicity, I want to share what I have learned about finding the causes for these experiences and how to repair them. The vast majority of the patients I treat do get better, and most are cured. We have learned a lot in the past twenty years, but we still have a long way to go. What we have learned thus far may help you find your path back to healing, and that is why I wrote this book.

Please read on.

PART 2

UNDERSTANDING THE CAUSES OF SENSITIVITY AND TOXICITY

It's All in a Good Cause

" It doesn't matter how much you drive around, you will never get to where you want to go if you don't have the right map. "

—Stanley Rosenberg,
from *Accessing the Healing Power of the Vagus Nerve*

Once we can wholeheartedly accept the reality of extreme sensitivity and/or toxicity in patients, the first obvious task is to seek to understand its cause. Throughout this book, I will emphasize that the single most important component of treatment is making a precise diagnosis. Simply labeling sensitivity as multiple chemical sensitivities (MCS) or accepting it as real, though reassuring to the patient, is only the first step.

Identifying what we are currently calling chronic inflammatory illness or environmentally acquired illness is just the beginning of a somewhat complicated diagnostic process. It begs the question: What, exactly, is triggering this illness?

Over the last ten years, we have learned quite a bit. A wide variety of toxic substances and infections can be the culprits. Any number of viruses, bacterial infections, and parasites have been found to be possible inciting causes. Any number of environmental toxins, of which there are a bewildering array, also have been noted to be potential causes. So where do we start?

In this part of the book, I hope to simplify this process while at the same time honoring its complexity. In my experience, the two major causes of overwhelming sensitivity are mold toxicity and Lyme disease (with its co-infections). While there are many, many other possibilities, nothing else comes close to these two entities as a probable cause. I would estimate that 80 percent of my sensitive patients are struggling with mold toxicity as the primary trigger, and 20 percent were triggered by Lyme disease, primarily *Bartonella*. A large percentage have both.

Both of these conditions profoundly weaken the immune system, which opens the door to all kinds of opportunistic infectious agents that have been dormant in the body and allows them to join the party. Although those microbes, most prominently species of *Mycoplasma* and *Chlamydia*, the Epstein-Barr virus, and human herpesvirus 6, may play a role here, rarely in my experience are any of those the primary problem. They may have contributed to an initial weakening of the immune system that predisposed that system to harboring molds or developing Lyme disease, but rushing to treat those microbes has not helped the majority of my patients.

The immune system is a bit unique in that usually it can wrestle with only one microbe at a time, so it prioritizes what it perceives as the most significant threat to the body and mobilizes the system to fight that threat, primarily. As I mentioned in the discussion of Karen's case back in Chapter 1, I call this identifying Public Enemy #1. *Even if other infectious agents are present, unless treatment is in sync with what the immune system has identified as its primary objective, treatment will be working at odds with that system.*

A common example is starting treatment with a focus on the Epstein-Barr virus (EBV). The vast majority of my patients test positive for having had exposure to EBV, but that does not necessarily mean it is currently active. Furthermore, the symptoms caused by EBV rarely provide anything like the complete picture presented by my patients' symptoms. Patients are often referred to me after being treated for EBV for several years with minimal benefit. While EBV may be present, that is not what the patients' symptoms are telling us is their Public Enemy #1.

It is imperative that we correctly identify the illness that *is* Public Enemy #1 so we can begin effective treatment that gets directly at the cause. I have found that once those triggering conditions, notably mold toxicity and Lyme disease, have been successfully treated, the immune system has recovered sufficiently to bring back under control all of those previously latent infections without requiring additional treatment. Occasionally, when the major conditions have been properly treated, we see some residual symptoms that inform us as to what still needs to be addressed, but these cases are relatively infrequent.

For this reason, I will emphasize mold toxicity and *Bartonella* infection as the two major sensitizing conditions that need to be diagnosed and treated as the primary task of the physician. This is not to say that there are not other conditions that can contribute to chronic inflammatory illnesses. However, while those other conditions can cause all kinds of problems, they are not what I call the *primary sensitizers.* They can, and do, make patients sick, but they are rarely the trigger that creates the out-of-control sensitivity or reactivity that makes life so difficult.

The underlying theme for all of these conditions, whether they are triggered by toxins or infectious agents, is inflammation.

Toxins and microbes stimulate the immune system to produce inflammatory cytokines, which are messengers used by the body to regulate the immune response. For reasons we are slowly beginning to understand, the body is unable to turn off this inflammatory process once it is stimulated by these specific causal elements. What begins as a normal, healthy immune response over time develops into an out-of-control chronic illness. Think of it as a warming campfire on a cold night that erupts into a devastating wildfire.

As our understanding has evolved, the nomenclature has changed: what used to be termed chronic fatigue syndrome (CFS)

> **The underlying theme for all of these conditions, whether they are triggered by toxins or infectious agents, is inflammation.**

is now known as environmentally acquired illness or chronic inflammatory illness. This change reflects the realization that what was once considered a psychological illness is clearly a physical one, with inflammation at its heart. This has been demonstrated and reported in the medical literature by Robert Naviaux, MD, and his team in a groundbreaking paper (described in Chapter 8) that details the biochemical—not mental—changes that occur in patients with CFS. Finally, the stigma of CFS as a psychological issue has been addressed, and we can move forward with more specific evaluations and treatments.

Although many other conditions have been linked to inflammation in patients with CFS and fibromyalgia, I have found that mold toxicity and *Bartonella* are the main culprits. If a physician is presented with such a complicated patient and does not look in that direction, I fear that little progress will be made, and the patient will get progressively worse and become even more sensitive.

This part of the book is devoted to understanding these primary causes and the important secondary contributing factors (triggered by mold toxicity and Lyme disease) of mast cell activation, porphyria, and carbon monoxide poisoning. Because I find that mold toxicity is far and away the most common trigger, I will start with that in Chapter 3 and spend a great deal of time explaining what it is and how to treat it. I will then move on to *Bartonella* in Chapter 4. Both mold toxicity and *Bartonella* infection often trigger the additionally complicating processes of mast cell activation, which is discussed in Chapter 5, and porphyria and carbon monoxide poisoning, discussed in Chapter 6. Though not caused by mold toxicity or Lyme disease, another important and largely unrecognized cause of an increase in sensitivity is carbon monoxide poisoning, which Chapter 6 will also bring to light. I will try to tie some of this material together in Chapter 7, "Where to Start?"

This field is so new and so complicated that there is no way I can cover every conceivable insult to the body that might create this sort of sensitivity reaction. Because I am focusing on four major subjects in this part of the book, I might be accused of oversimplifying this process, and yes, I am guilty of that. By doing so, however, I hope to draw attention to what I have found to be the most important areas to delve into that have resulted in the best results. I hope that other physicians will be able to help more and more of these suffering individuals. I anticipate that we will learn a great deal more about both causes and treatments over the next few years.

CHAPTER 3

Mold Toxicity

There's Mold in Them Thar Hills

When it comes to the most sensitive patients, I have emphasized (and will continue to emphasize) that mold toxicity is the single most common contributing cause of that sensitivity. Accordingly, this chapter is the longest and most detailed one in this book. It is my hope that this information will be valuable for healthcare providers and patients alike. At this time, other than my e-book *Mold and Mycotoxins,* there isn't much available for either group to read that gives a complete perspective of what is known about this subject, and it is my intention to translate that material into a form that is relevant for especially sensitive patients.

Mold toxicity is far more common than is currently recognized. Those who are knowledgeable in this field estimate that millions of people are wrestling with this problem but are entirely unaware of its existence. Mold toxicity goes so unrecognized by most medical practitioners that a patient bringing it up as a possible diagnosis is usually met with a blank stare or, worse, incredulity. This requires patients to become their own healthcare advocates because they cannot rely on their physicians to make a diagnosis of which those physicians are unaware.

Despite its prevalence and importance, this field was barely recognized until 2005, when Ritchie Shoemaker, MD, published his pioneering book *Mold Warriors.* Our current knowledge, though helpful, is still limited by a lack of scientific studies that would inform us how to work in this field with more precision. *What we don't know far outweighs what we do know.* But what we do know is extremely useful and allows us to help the myriad of patients who would remain ill without intervention.

The good news is that some pioneering physicians have grasped the significance of accepting mold toxicity as a valuable diagnosis, and we are making rapid improvements in our abilities to treat it. I anticipate an explosion of increasingly scientific new information in the near future, so please view this chapter as simply the best introduction I can provide today. It is subject to change at a moment's notice but is still useful as a starting point for learning.

What Is Mold Toxicity?

A *toxin* is simply a poison. Microbes in our bodies can produce these poisons that make us sick, and then sicker—a phenomenon referred to as *biotoxicity*. It is fairly obvious that microbes (bacteria, viruses, fungi, parasites, and other infectious agents) can make us sick by causing infection, and it is equally clear that treating for these disease-causing microbes will make us better. But what is just now becoming clear is that not only do these microbes make us sick by causing infection, but some of them can, in the process of being killed by our immune systems, release toxins into our bodies that make us even sicker. Because these toxins are made by microbes, which are living biological systems, they are called *biotoxins*. This term distinguishes them from other types of toxins, such as heavy metal toxins or synthetic toxins, that are not made by living systems.

Doctors are beginning to appreciate that some of these toxins remain in the body and are difficult to excrete or destroy. Thus, they accumulate. To make things a little more complicated, many of these toxins actually *compromise the body's ability to detoxify* so that as they accumulate, they become increasingly more difficult to eliminate, creating a vicious downward spiral.

The body's natural method of processing toxins is to concentrate them in the body's main organ of detoxification: the liver. Once the biotoxins congregate in the liver, they are bound to bile and sent out into the gastrointestinal tract for release in the stool. However, the body's natural recycling system, called *enterohepatic circulation,* recirculates the bile when it reaches the small intestine. So the toxins, still attached to that bile secreted by the gallbladder, go back to the liver rather than leaving the gastrointestinal system and thus accumulate in the body. The purpose of enterohepatic circulation is simply to conserve bile, but in this instance, this conservation effort backfires.

Hence, even when the immune system kills the invading microbes, the remaining toxins can continue to plague us with their harmful effects.

Not everyone is subject to this problem. In about 75 percent of the population, the immune system recognizes these toxins and uses its defense mechanisms to destroy them. Unfortunately, the remaining 25 percent are genetically unable to make antibodies to these toxins. These are the patients who are prone to becoming progressively sicker as toxins accumulate in their bodies.

Which specific microbes are we talking about? The most common offenders are molds, including the black mold *Stachybotrys*, *Aspergillus* (which can also be black, so don't let the color of the mold that is growing limit your understanding of which molds may be present), *Penicillium*, *Fusarium*, *Chaetomium*, *Alternaria*, and *Wallemia*. Harmful toxins can also be produced by microbes other than mold, such as certain viruses, especially those in the herpes family, such as the Epstein-Barr virus, which is the agent of mononucleosis, as well as by Cytomegalovirus (CMV) and human herpesvirus 6 (HHV-6), which is the agent of roseola infection. A particularly toxic strain of HHV-6 has been linked to chronic fatigue and fibromyalgia. Lyme bacteria can also produce toxins, as can the common co-infections of Lyme, such as *Bartonella* and *Babesia,* along with species of *Mycoplasma* and *Chlamydia.* The atypical pneumonias (walking pneumonia) and Gulf War syndrome are examples of these last two infective agents. In this chapter, I will focus on mold.

Mold toxicity is not new. It is described in some detail in the Old Testament in Leviticus 14, which warns that if a man's house is contaminated with plagues, mold, and leprosy, "he shall have the inside of the house scraped all around, and the plaster that they scrape off they shall pour out in an unclean place outside the city." It has been known for decades that mold toxicity can be a significant medical problem, yet relatively few studies have been published on this topic.

During the 1970s and 1980s, the United States Defense Department grew concerned about the possibility of mycotoxins being used for biological warfare and did some useful research on how to *bind,* or rid the body of, some of these substances. Despite these efforts, the subject of mold toxicity was almost unknown to most physicians at that time. In the 1990s, a few scattered case reports were put forth in the medical literature, mostly out of Scandinavia. A landmark paper from the Mayo Clinic in 1999 suggested for the first time that the cause of chronic sinus infections might not be bacterial infection, but rather fungal infection, and the researchers demonstrated improvements when patients' sinuses were treated with intranasal amphotericin B. Several years later, in 2003, Dr. Michael Gray and his research team published several papers that clearly demonstrated both immune and nervous system dysfunction in more than 200 patients with mold exposure from water-damaged buildings, and the team proposed that this condition be called *mixed mold mycotoxicosis*. Still, the medical profession did not yet realize how commonly this occurred and how ill patients who were exposed to mold could become.

Ritchie Shoemaker was a family physician in Pokomoke, Maryland, who saw an outbreak of unusual symptoms in his patients when their local environment, the Chesapeake Bay, saw the death of thousands of fish in 1997. Dr. Shoemaker, working with local scientists, tracked down the cause of this illness to the presence of toxins created by the dinoflagellate *Pfiesteria*. He also discovered that these toxins could be bound by the medication cholestyramine, which enabled his patients to recover. He later found similar microbes *(Cylindrospermopsis)* in algae-bearing lakes in Florida and in long-standing cases of ciguatera (a food-borne illness that occurs as a result of eating certain contaminated seafood). He realized that, along with mold exposure, all of these illnesses shared a pattern that was consistent and treatable. We suspect that other microbes will be implicated as our understanding of this problem deepens.

I would like to emphasize (as Dr. Shoemaker correctly points out) that when we talk about mold toxicity, we are including not only the presence of mold itself but also mold spore fragments, volatile organic compounds (VOCs), and the microorganisms *Actinomycetes* and *Mycobacteria,* which are usually found along with toxic mold species, as well as beta glucans, hemolysins, mannans, and proteinases, all of which contribute to the toxicity. These irritating components of "mold" exposure require us to think more inclusively rather than to consider the fungal parts of the mold itself as the only causative element present.

Symptoms of Mold Toxicity

We have a special word in medicine that describes a symptom so specific that the mere mention of it is sufficient, at times, to make a clear diagnosis. That word is *pathognomonic* ("path-og-no-mon-ic"). It is particularly relevant to mold toxicity in that quite a few symptoms are so particular to patients with mold toxicity that we immediately gravitate toward that diagnosis. Specifically, they are:

- Electric shock sensations
- Ice pick–like pains
- Vibrating or pulsing sensations running up and down the spinal cord (sometimes confined to specific segments of the spine)

If these descriptions are associated with many of the symptoms listed below, particularly muscle weakness, numbness and tingling in different parts of the body, disequilibrium, dizziness, severe anxiety and depression with fatigue and cognitive impairment, joint and muscle pain, headaches, gastrointestinal symptoms, and chest tightness and pain, the possibility of mold toxicity should immediately jump to mind.

Unfortunately, patients with these symptoms often find it difficult to put their experiences into words, and they have learned that when they try to do so, eyebrows will be raised, especially if their doctors are not familiar with mold toxicity. They find themselves instantly judged as having a psychological condition, and their doctors stop listening to their descriptions. Medical arrogance, alas, is such that if a patient describes something a physician has not yet studied, those symptoms are "in the patient's head" until proven otherwise. Mold patients quickly learn to be very careful about how they describe their symptoms, and it often takes them time to feel safe enough to fully open up.

So, when I hear even the faintest hint that a patient is experiencing some of these symptoms, I ask open-ended questions that are meant to convey the following:

"Yes, I have heard these descriptions before. You are not alone in this experience, and these symptoms are meaningful to me. They tell me not only that this is *not* in your head, but that there is a physical cause that we can diagnose and treat."

If they can be elicited from understandably reluctant patients, descriptions of these unusual sensations (patients often characterize them as "weird") can be key in diagnosing mold toxicity.

For many years, when patients would express a litany of unusual symptoms, the missing diagnosis was Lyme disease with its co-infections. This is why, for several *decades,* it would take five to fifteen years or more for physicians to diagnose Lyme disease in patients who were struggling with it. Simply put, it was the diagnosis of desperation—the last thing on any physician's list—and so controversial that doctors were reluctant to entertain Lyme as a possibility. But in 2013, the U.S. Centers for Disease Control (CDC) released a report stating that around 300,000 new cases of Lyme disease are discovered in Americans *every year*. This official sanctioning has moved Lyme up the list and has slowly but surely made it possible for physicians to diagnose it much earlier in its clinical course.

This makes mold toxicity the new replacement for the most under-appreciated, under-recognized illness that I can think of. As was the case with Lyme disease, most physicians are unaware of the symptoms and clinical course of mold toxicity, and once again medical arrogance rears its ugly head such that most patients with mold toxicity symptoms are summarily dismissed as being psychologically disturbed. The key here is that having a good grasp of the symptoms that delineate mold toxicity may help prevent years of suffering.

So, what are the symptoms of mold toxicity, or what can be thought of more globally as biotoxicity? Note that mold toxicity is one of the most common examples of toxicity caused by biological organisms. Bacterial infections, such as Lyme and its co-infections, create toxicity by releasing their cell contents when they die or are killed, as do a number of rarer biological exposures, such as *Pfiesteria, Cylindrospermopsis,* ciguatera poisoning, and brown recluse spider bites. Biotoxicity is a larger category that includes mold toxicity, but mold toxicity, as the major player here, will be our focus.

Let me begin by going over the full spectrum of symptoms that are possible with mold toxicity. Obviously, not every patient will have every symptom listed here, but the wide array of symptoms helps us understand how complicated this issue can get and why mold (like Lyme) can be called a "great masquerader."

Patients with this problem experience a surprisingly wide range of effects relating to many different organ systems. When this confusing array of symptoms is taken out of context or is not understood as representing the many manifestations of biotoxicity, it is easy to see how both patients and physicians might mistakenly think that the problem is "all in their heads."

Common Symptoms of Mold Toxicity

Fatigue

Weakness

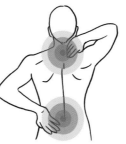

Muscle aches and cramps

Headaches

Sensitivity to bright light

Unusual pains

Abdominal pain, nausea, diarrhea

Chronic sinus congestion

Coughing, chest pain, shortness of breath

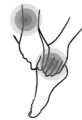

Joint pain with morning stiffness

Cognitive impairment

Skin sensitivity to light touch

Numbness and tingling

Sensitivity to electrical shocks

Metallic taste in the mouth

Excessive thirst

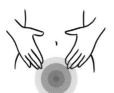

Menorrhagia

Disequilibrium and dizziness

Other symptoms:

- Appetite swings and weight gain
- Anxiety and depression
- Mood swings
- Night sweats
- Frequent urination
- Body temperature dysregulation
- Impotence
- Odd tics and spasms and seizure-like events

I would like to reiterate that several of these symptoms are somewhat unique, and if they occur together, they are a strong indicator for doctors to look for mold toxicity as a cause of illness. These are the painful "ice pick" or "lightning bolt" sensations, unusual areas of numbness and tingling (which neurologists are likely to deem psychological because they do not correspond to the typical nerve distribution patterns), odd tics and spasms and seizure-like events, and disequilibrium and dizziness that cannot be easily diagnosed neurologically (a neurological diagnosis of "atypical" multiple sclerosis, Parkinson's, Alzheimer's, or pseudoseizures is often a tip-off). If these symptoms are reported in conjunction with fatigue, cognitive impairment, intense anxiety or depression, and odd buzzing or "tremor" sensations in the body, physicians should suspect mold toxicity.

While many of these symptoms sound a lot like the symptoms of fibromyalgia, chronic fatigue syndrome, or depression, that's exactly how this illness may manifest. In a study published by Joseph Brewer, MD, and colleagues in 2013, 93 percent of a group of 112 patients with chronic fatigue syndrome were noted to have elevated levels of mold toxins in their urine! In children, the symptoms look a lot like ADHD. In fact, many patients with these particular diagnoses may have unrecognized biotoxicity as either a component or the direct cause of their illness, and most of them would benefit greatly from an evaluation for mold toxicity.

How Mold Toxins Get into the Body

Molds make toxins called *mycotoxins,* primarily to keep other molds out of their ecological niche. (John Banta will discuss this in more detail in Appendix A.) Mycotoxins can enter our bodies if we inhale the spores into our lungs, absorb the toxins through our skin by direct touch, or ingest them in the form of contaminated food. Certain foods, such as dried fruits, aged cheeses, mushrooms, overripe fruits and vegetables, beer, wine, wine vinegar, and processed meats, are known to contain minute amounts of specific mycotoxins, but the expert consensus is that inhaled spores are the primary source of mold toxicity for the vast majority of patients. This means that checking a person's living areas (home, workplace, and/or car) is the main focus for clarifying ongoing exposure. It also means that patients who have been identified as having mold toxicity should not touch or attempt to clean up visible mold, as doing so could result in an immediate worsening of their condition.

The Biotoxin Pathway

If you are not a healthcare provider, you may want to skip this section, as it is fairly technical. You can still understand the basics of mold diagnosis and treatment if you do not read this section, or if you merely skim it. What I have written is, believe it or not, a simplification, but there is no way to make these interactive biochemical reactions much less complicated. I encourage you to check out Dr. Shoemaker's excellent Biotoxin Pathway diagram on his website, www.survivingmold.com/diagnosis/the-biotoxin-pathway, so that you can follow along more easily.

By 2005, Dr. Shoemaker had put together a clear understanding of how mold toxins interfere with human body chemistry. In *Mold Warriors,* he calls this the Biotoxin Pathway. It is beyond the scope of this book to cover this chemistry in detail. It is my intention to convey the most up-to-date information about how to diagnose and treat mold toxicity. In the service of explaining how this subject has evolved, however, I would like to provide a somewhat oversimplified review of this material, which Dr. Shoemaker and I jokingly named "Shoemaker for Dummies" many years ago.

Mold toxins are unique. As noted in Chapter 2, these relatively small molecules are called *ionophores* because one end is *lipophilic,* meaning that it dissolves easily in lipids (fats), and the other end is *hydrophilic,* meaning that it dissolves readily in water. While this might not, at first glance, seem important, the basic premise here is that oil and water do not mix. If you shake them up, they will quickly separate, leaving a clear line demarcating them. Because mold toxins have the ability to dissolve into fats at one end and into water at the other, they essentially can move through any body tissue at will, as well as through all membranes (which usually can regulate which molecules can move through them and which cannot), insidiously forcing a wide variety of tissues to react to their toxic qualities. This limits the processes that that body can utilize to eliminate these toxins.

So, let's walk through the basic steps of this pathway.

> **Mycotoxins can be inhaled, eaten, touched, or, as we now know, made by molds that have colonized a body (most commonly the sinus and gut area) previously exposed to molds.**

First, the mycotoxins enter the body: they can be inhaled, eaten, touched, or, as we now know, made by molds that have colonized a body (most commonly the sinus and gut area) previously exposed to molds.

Once the toxins are inside the body, the body attempts to deal with them via an orchestrated process. Mycotoxins initially enter a fat cell, where toxins are stored by binding with a Toll receptor on the *outside* of that cell. This binding is followed by the release of a second messenger, which now moves *into* the cell, binding to and activating a third messenger called NF-kB ("NF kappa B"). NF-kB moves into the nucleus of the cell, where it turns on gene transcription for a DNA segment that codes for the production of cytokines. *Cytokines* are immune messengers that in this situation stimulate a profound inflammatory response.

To put this in even simpler language, the mycotoxin can enter a cell and set off a severe inflammatory process. As mentioned earlier, most people are not susceptible to this reaction because they are genetically blessed with the ability to bind and get rid of this toxin before it causes all this damage. Unfortunately, 25 percent of people lack this gene and are predisposed to being affected by mold toxins; these are the patients we see with mold toxicity. There is nothing psychological about this genetic predisposition to illness caused by mold toxins, but because one or two individuals in a home or workplace can be profoundly affected by mold toxins while others in the same environment are perfectly fine, you can understand how people might assume (incorrectly) that this reaction must be psychological.

Coming back to our discussion of the Biotoxin Pathway, this flood of cytokines blocks the body's leptin receptors. *Leptins* are important hormonal regulators of satiety (when you feel like your hunger is satisfied and can stop eating) made by fat cells. With the leptin receptors damaged by the excess cytokines, the fat cells crank out more and more leptin in a vain effort to override this blockage and get the system working properly again. This creates *leptin resistance.* Even more important, the hypothalamus can no longer make adequate amounts of MSH (alpha-Melanocyte-stimulating hormone) and VIP (vasoactive intestinal polypeptide), which are critical regulators of neurological, immunological, and endocrine function (and more), as I will get into shortly. The important point here is that the mycotoxins can, by causing this uncontrolled inflammatory release of cytokines, *directly impact* immune regulation, hormonal balance, and brain and nerve function, leading to a confusing array of symptoms.

Let's spend a little more time on cytokines because they are so crucial to this pathophysiological process. Cytokines are special proteins (there are lots of them) that are responsible for both healthy and inflammatory immune processes. For example, if you are infected by a nasty virus, your first symptoms may include fatigue, joint and muscle aches and pains, fever and chills, and brain fog. It is often assumed that these are the symptoms of the virus in the body, but they are not. These symptoms are created by the immune system's reaction to the virus: in an attempt to kill the virus, your immune system releases a flood of cytokines, and it is this flood that causes your symptoms. You see, your immune system is recognizing a foreign invader (in this scenario, a virus, but it can also be a bacterium or other microbe) and is setting off an alarm reaction to mobilize cytokines to help deal with this threat. This triggers an immune response to recruit additional immune cells, like macrophages and lymphocytes (both are types of white blood cells), that enter the battlefield and make even more cytokines in an effort to complete the job. When the immune system detects that the invaders are gone, it is supposed to send out a signal that ends this reaction. But if it is a persistent toxin triggering this process and you are among the 25 percent of individuals who cannot readily get rid of these toxins, this out-of-control inflammatory process *persists.* This is the essence of the problem. We need to get the inciting toxins out of the body, quiet down the immune system, and deal with the fallout, which I will discuss below.

If a person's immune system can make antibodies to the toxins, he or she won't get sick, or the symptoms will be minor. The key to this process is called *antigen recognition.* This is the immune system saying, "Ah, yes, I have seen this microbe or toxin before, and I know exactly what to do with it." Without going into great detail about antigen recognition (which can be very complicated), it does require an elaborate dance, if you will, in which the foreign antigens (microbes or toxins) must be packaged in a particular form and presented to special dendritic cells, which in turn present these antigens to special lymphocytes for more processing, finally getting the B lymphocytes to make antibodies specific for these proteins. *Toxins can disrupt almost every aspect of this presentation process,* preventing the correct antibody formation and preventing the body from being able to clear antigens (which are relentlessly stimulating the formulation of cytokines). This creates a persistent, dysregulated response of the innate immune system.

Without antibodies to directly clear toxins, the only mechanism left to eliminate them is the organic anion transport system in the small bile ducts of the liver, which secretes the toxins into the bile, binding the toxins to the bile that eventually flows into the small intestine. I referred to this earlier, but it is worth repeating: unfortunately, the toxins don't stay in the intestine for elimination but, still bound to the bile, are reabsorbed through a process called *enterohepatic circulation* before they reach the colon and return to the liver and gallbladder. Grasping this is vital to the recognition that we must find more efficient binders that will assist these toxins out of the body through the gastrointestinal tract.

With these basics in mind, let's go back to the Biotoxin Pathway to learn more about the biochemical markers (specific substances that can be measured in the blood) that Dr. Shoemaker has discovered that will shed additional light on what is happening, how we can measure it, and what we can do about it.

One component of this overstimulated innate immune system is called *complement*. Serum measurements of two components of this system, c3a and c4a, have proven valuable in evaluating the immune system's ability to turn itself off. Although many physicians routinely use two measures of inflammation, called the erythrocyte sedimentation rate (or "sed rate") and C-reactive protein, these lab tests have not proven at all useful for measuring the state of inflammation in patients who are sick with biotoxin illnesses. The kind of inflammation we are seeing in these patients does not seem to be reflected by those tests. However, patients often are relieved to learn that they have markedly elevated c4a, finally confirming what they have been perceiving as true. (They almost always say, "I feel so inflamed.") Another biochemical marker that has proven useful in demonstrating the presence of prolonged inflammation is the blood test TGF beta-1 (transforming growth factor beta-1). Please do not be intimidated by the array of acronyms being presented here; they are just the names of substances that may be helpful to follow.

MMP-9 (matrix metalloproteinase-9) can be another valuable measurement. Matrix metalloproteinases are protein-digesting enzymes that dissolve molecules in the tissues beneath the cell membranes in blood vessel walls. They eat through these basement membranes and allow inflammatory compounds to go from the bloodstream into tissues such as the brain, joints, nerves, and lungs, where they can do additional damage. Detecting an elevated level of MMP-9 in the blood may be helpful both in diagnosing biotoxin illness (it is especially elevated in patients with mold toxicity and bartonellosis) and in determining how far it has progressed.

An additionally useful measurement is for a substance called VEGF (pronounced "veg-F"), short for vascular endothelial growth factor. Another result of the outpouring of cytokines detailed above is that when endothelial cells (the cells that line blood vessels) are exposed to these cytokines, the cytokines may bind to receptors on those blood vessels and cause the endothelial cells to release the "glues" that hold the linings of the blood vessels in place. These glues (adhesins and integrins, technically) can bind to circulating white blood cells and hold *them* in place at the receptor site. This creates blockages in the small capillaries, narrowing them and preventing normal blood flow. This means that the tissues downstream from the blockages do not get adequate circulation and oxygen. When the body senses this (lack of oxygen is a priority for the body to monitor), it releases a gene controller called HIF (hypoxia inducible factor), which produces VEGF to stimulate blood vessel growth to circumvent the blockage. This works well as a healing mechanism in healthy people, but in patients with mold toxicity, high cytokine levels suppress VEGF production, *and blood flow through the capillaries remains compromised.* This both causes and adds to the symptoms of fatigue, cognitive impairment, persistent myalgias, and poor recovery from even mild exercise. Mitochondrial function is impaired. A low VEGF measurement correlates with these physiological events.

While these biochemical changes are important, perhaps the most central effects created by the dysregulated outpouring of cytokines are those on the hypothalamus. The hypothalamus is the part of the brain that controls the release of most hormones; it regulates the pituitary gland, which is known as the "master gland." These hormonal imbalances are *directly* caused by the effect of toxins on the hypothalamus. (See Chapter 11 for a detailed account of these hormonal imbalances.) Excessive cytokines bind to receptors in the middle of the hypothalamus for what is called the "super molecule": proopiomelanocortin. Despite its intimidating name, proopiomelanocortin is simply a molecule that is broken down into three important components: *pro* means that the molecule is a precursor to the other molecules that are activated; *opio* refers to endorphins, which are our natural "opioids" for relieving pain as it arises; *melano* refers to MSH (alpha-Melanocyte-stimulating hormone), which I referred to above and will discuss in more detail soon; and *cortin* refers to ACTH, or adrenocorticotropic hormone, which stimulates the pituitary gland to make the essential adrenal hormones, including cortisone.

Coming back to the overproduction and unregulated production of cytokines induced by mold toxins, we return to the particular importance of MSH. In a healthy person, leptins bind to receptors that activate and produce MSH. In a biotoxic patient, however, the flood of cytokines blocks the leptin receptors. As a result of leptin resistance, not enough MSH is produced, affecting the production of adrenal hormones, sex hormones, thyroid hormones, and anti-diuretic hormone (ADH), which regulates how the kidneys regulate water and fluid levels in the body, increasing pain (from the decrease in endorphins), reducing sleep (from the decrease in melatonin), creating a leaky gut, which in turn increases the risk of autoimmune disease, and weakening the immune system's ability to respond to underlying infections. You can immediately see how many organ systems can be affected and how profound and widespread those effects can be. Understanding this process allows us to make sense of the remarkable variations in symptoms presented by mold-toxic patients.

In addition, MSH regulates protective cytokine responses in the skin, lungs, gastrointestinal tract, and nasal mucous membranes; the loss of MSH makes those areas of defense more vulnerable to invasion and colonization. The dysregulation of endorphins predis-poses people to illnesses such as fibromyalgia and irritable bowel syndrome (IBS), where the decrease in endorphin production sensi-tizes nerves to incoming stimuli. In women, vulvar pain and interstitial cystitis are common. ADH dysregulation means that the kidneys can-not do their job properly and cannot prevent the loss of free water. Patients urinate more often and become more easily dehydrated. (They often bring large quantities of water with them to clinic visits.) Dr. Shoemaker has proposed that their sweat glands dump extra salt through the skin, predisposing them to shocks of static electricity, which are common in this patient population.

MSH deficiency (which can be measured with a blood test) coupled with weakened immune defenses makes a person prone to being colonized by a subset of staph infection called MARCoNS (Multiple Antibiotic-Resistant Coagulase Negative Staphylococci). This is a *colonization,* not an infection, but an important one. The dis-tinction here is that this bacterial colonization does not cause symp-toms (such as nasal congestion or sinus infection) directly, but its physiological effects may contribute to a worsening of mold toxicity. Dr. Shoemaker notes that these organisms make hemolysins, which destroy red blood cells. The body views the debris of these dam-aged red blood cells as a foreign invader to which it must respond by making more cytokines, causing an increase in the inflammatory response. Of even more relevance, MARCoNS can make exotoxin A, which can split MSH, destroying its function. Additionally, MARCoNS can make thick layers of biofilm that surround these bacteria, making it difficult for the immune system and treatments to get at the offending organisms. You can see that looking for MARCoNS (which is done with a nasal swab sent to a lab capable of evaluating such swabs) and treating it becomes an integral part of the treatment of mold toxicity.

To complete this "basic" discussion of the Biotoxin Pathway, I need to discuss the important role of VIP. The unregulated outpouring of cytokines to which I keep referring also shuts down the production of VIP. Like the other markers noted here, which were discovered and researched by Dr. Shoemaker, VIP can be measured by a blood test, but to be accurate, it needs to be sent to one specific facility, ARUP Laboratories, for evaluation.

VIP is a neuropeptide that functions as a neuromodulator and neurotransmitter. Early studies found that it is a potent vasodilator that regulates smooth muscle activity, epithelial cell secretions, and blood flow in the intestinal tract. As a chemical messenger, it functions as a neurohormone. Interestingly, it is produced in the gut, pancreas, and brain from the hypothalamus region.

Of particular importance is the recognition that VIP plays a significant role in immunomodulation. Recall that one of the particularly deleterious effects of mold toxins is to prevent the body from releasing VIP (which Dr. Shoemaker has reported in the majority of his patients, and which I can confirm). A groundbreaking paper by Mario Delgado and colleague notes that VIP, "released in the lymphoid organs by the innervation and activated immune cells, modulates the function of inflammatory cells." Translating complicated immunology into English, this means that if the body cannot make VIP, it cannot control inflammation, which therefore rages out of control. This out-of-control inflammation is a central feature of mold toxicity, Lyme disease and its co-infections, and all of the other chronic illnesses we are seeing. Chronic inflammation is the cardinal issue creating the myriad of symptoms that complicated patients are wrestling with. So, not enough VIP means no control of inflammation. I will go into more specifics of how to work with VIP in the treatment section of this chapter, beginning on page 62.

> **Chronic inflammation is the cardinal issue creating the myriad of symptoms that complicated patients are wrestling with.**

Diagnosing Mold Toxicity

The first order of business is to even *consider* the diagnosis of mold toxicity. Until I read Dr. Shoemaker's books and studied with him, I had never thought to ask my patients about exposure to mold or other infectious agents. Now I routinely inquire about possible exposure to mold at home or at work. Have the patients' homes sustained any water damage or leaks, including the roof, basement, crawlspace, and walls? Do they notice a musty smell or see any mold? The materials that are visible are often just the tip of the iceberg. Mold can grow inside walls and get into the heating and cooling system of a house, sending spores spewing over the entire residence. If you are wondering how this happens, be aware that sheet rock paper is made from processed tree bark, which is loaded with mold spores. To bring those spores to life, just add water. (See Appendix A by John Banta for more information about mold growth.)

How about at work? Do co-workers have similar symptoms or complaints? And what about the air conditioning system in your car? As I have explored this area, I have found it surprising how many of my sick patients report exposures that they had never suspected as being relevant to their illness.

Keep in mind that mold toxins can remain in the body for long periods, so the mold exposure might have occurred in a previous residence or workplace. The toxins may be present in the body even several *years* after leaving a moldy home or workplace. In fact, for most of my sickest patients, the exposure occurred many years ago, often when their immune systems were already compromised. The mold may have colonized within their bodies to produce mycotoxins steadily, on an ongoing basis, even if there is no mold in their current environments.

Some physicians dismiss this subject, pointing out that in the natural world, we are surrounded by molds, so why make a big deal out of them? It is certainly true that nature is filled with molds—thousands of species of every shape and description. In fact, the reason mold species make toxins is not to damage *us,* but to keep other species of mold at bay as the species compete for their own piece of real estate in the environment. It is a system of checks and balances, with each species holding the ones nearby at arm's (spore's) reach. However, when a species of mold can grow inside a dampened wall without competition, it grows unchecked, and as it reproduces at will, it can make significantly more mycotoxins and spores. While most of these species are relatively harmless to humans, several, including *Stachybotrys, Penicillium, Wallemia, Chaetomium, Fusarium, Alternaria,* and *Aspergillus,* are capable of producing a toxin that can make us quite ill. *Stachybotrys,* or black mold, is the most well-known, but the others can be even more dangerous, and some of them also appear black to the naked eye.

I can't emphasize enough that the most important component of diagnosing mold toxicity is to *consider that possibility* when a patient describes symptoms that are across-the-board complicated.

MAUREEN'S STORY

Maureen was thirty-eight when I saw her for the first time in March 2006. She was one of my first mold patients. Her symptoms had begun three years earlier, when she was working in a building with a very leaky roof. She had experienced several traumatic events at the same time, including a difficult divorce, which complicated both our evaluation and her perceptions of what might have triggered her illness. In August 2005, Maureen had begun to violently throw up on a daily basis, and this went on for months. She saw multiple physicians, including several gastrointestinal specialists, and underwent extensive testing that showed no obvious cause for her symptoms. Various physicians diagnosed her as having bipolar disorder, irritable bowel syndrome, and hypochondriasis (meaning it was all in her head, or psychosomatic).

Since her initial exposure to the mold, Maureen had gained 35 pounds and had developed sensitivities to all sorts of chemicals and perfumes (which we call multiple chemical sensitivities, or MCS)—a common scenario following untreated biotoxin exposure. She continued to experience daily nausea and vomiting, regular headaches, and an inability to think clearly. She described her poor decision making in simple terms: "It's like I'm inept." She also reported blurred vision and bouts of chest tightness with wheezing.

When Maureen came to see me, she had just lost her job of fourteen years, at which she had previously been quite competent. She was trying to live a normal life and raise her three children, but she could hardly function. Her energy level was very low, and she reported having difficulty staying asleep and waking up unrefreshed each morning. She described joint pain in her neck, shoulders, and upper back, and she displayed the classical irritable bowel symptoms of gas, bloating, distention, and diarrhea alternating with constipation.

When I tested her vision with the functional acuity contrast test (FACT), the results strongly suggested biotoxin exposure. Her DHEA test was a little low, as was her progesterone level. I started her on a natural progesterone cream, ¼ teaspoon daily, applied to her skin. I also put her on a small dose of DHEA, and we started cholestyramine, one scoop mixed with water three times daily.

When Maureen returned to my office just six weeks later, she reported that she was 80 percent better. Her energy had markedly improved despite the fact that she was still living in her home, which had obvious mold issues. She had a new job and was successfully working again, but she noticed that when she missed even a single dose of cholestyramine, she became lethargic again. She planned to remediate her home as soon as she could, and she was thrilled to have her life back; she was no longer plagued by nausea, vomiting, and chest and joint pain.

Parenthetically, she was so impressed with her improvement over just a short period that she bought a copy of Dr. Shoemaker's book Mold Warriors *to give to her family physician, who had essentially told her that she was a hypochondriac. He refused to even look at the book or to take Maureen's story seriously.*

Most of my mold-toxic patients have had experiences similar to Maureen's. Their symptoms are usually misunderstood as representing psychological imbalances, particularly depression and anxiety, but they do not respond well to antidepressants or anti-anxiety medications. That should be the tip-off to look deeper for the cause of these symptoms. Recall that 25 percent of the population is genetically unable to make the antibodies needed to deal with these toxins. These patients may live in homes with others who have no symptoms or work in mold-laden environments where others are not sick. Family and coworkers therefore assume that because only one person is sick, it is in that individual's head. These patients feel like no one is listening to them, including their physicians. The trouble is that no one *is* listening. Being ignored adds even more to the burden of suffering, often tearing apart families and friendships, the stress of which makes the patients even sicker. Now they are sick *and* depressed, but we can clearly see that sickness came first. It is difficult to fight a biotoxin illness, and even more difficult when no one believes that you are really sick.

The good news, however, is that we are beginning to understand the symptoms, and better diagnostic tools and more specific treatments are becoming available. Newer, more precise tests now exist for specific toxins, allowing us to take a more dynamic approach.

The simplest screening is a visual test called the functional acuity contrast test (FACT), also known as a visual contrast sensitivity (VCS) test. Ophthalmologists have been using this well-established tool for many years. The patient looks at a series of alternating white and black lines of decreasing clarity. There are five columns labeled A through E. Everyone can see A and B; the actual test is represented by columns C, D, and E. If the patient has difficulty seeing or is unable to see the lines in those columns, this indicates poor retinal function, which has been closely linked to biotoxicity. As shown in Figure 3.1, Maureen could not see the lines in column E at all and was not able to see the lines in columns C and D very well.

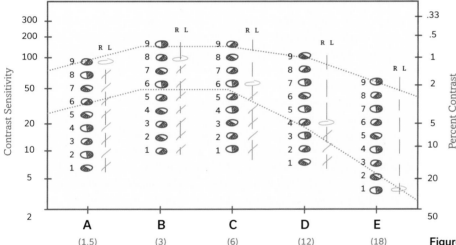

Figure 3.1. The results of Maureen's visual contrast test indicated the presence of mold toxicity.

If you suspect mold toxicity, I encourage you to go to Dr. Shoemaker's website, www.survivingmold.com, to take this test online. While many, if not most, mold-toxic patients fail this test, quite a few with strongly positive urine tests for mold toxins have normal FACT tests, so although this is an inexpensive screening option, that's all it is. If mold toxicity is suspected, we must dig deeper.

Dr. Shoemaker has discovered a connection between biotoxicity and a series of biochemical tests that show a patient is experiencing an inflammatory reaction to toxin. He has clearly demonstrated that a treated patient, when placed in a moldy environment, will have noticeably elevated levels of inflammatory markers (measured by blood tests), which return to normal after the patient is removed from the moldy environment and resumes treatment. These tests reflect an understanding of the Biotoxin Pathway.

Using readily accessible laboratories such as Quest Diagnostics and LabCorp, we can measure c4a, TGF-beta-1, MSH, VIP, VEGF, MMP-9, and leptins, giving us a snapshot of how a patient's immune system may have been impacted by mold toxicity. Markedly elevated levels of c4a, TGF-beta-1, MMP-9, and leptins and low levels of MSH, VIP, and VEGF all point us in that direction.

Figure 3.2. Normal ranges of inflammatory markers set by Dr. Shoemaker, as measured by blood tests. Markedly elevated levels of these markers indicate the possibility of mold toxicity.

VIP	23–63 (should be performed by ARUP Laboratories only)
MSH	35–81
MMP-9	85–332
c4a	0–2380
TGF-beta-1	< 2380
Leptin	Male: 0.5–13.8; Female: 1.1–27.5

The majority of the patients I see have consulted with conventional medical practitioners and usually have been told, "Your lab tests are normal." By not ordering the correct, more specialized tests and assuming that normal testing procedures will pick up anything of note (a far-reaching and incorrect assumption), many physicians are convinced that the patient is delusional in his or her description of symptoms and that a referral to a psychiatrist is imperative. When the correct tests are ordered, however, patients can see obvious abnormalities in their results. I often hear comments like, "I *knew* I was inflamed, and I *knew* something was wrong with me. Finally!"

Though I find these tests useful, they have not proven to be as specific as I had hoped for nailing down the diagnosis of mold toxicity. Other inflammatory conditions, including *Bartonella* infections in particular, but also Lyme disease, *Babesia, Mycoplasma,* and *Chlamydia* species (among others), can affect the results of these tests in a similar way. When a patient has several of these conditions, teasing the diagnoses apart becomes more difficult. We can confirm the presence of persistent inflammation, but not its cause for that particular patient.

For many years, the Shoemaker labs were the best, and sometimes the only, diagnostic tools we had, and as such were quite valuable. However, we now have the ability to measure specific mycotoxin species in urine, nasal washings, and blood, which allows us to diagnose with much more specificity and precision. The first lab to make this testing available was RealTime Laboratories in Houston, Texas. RealTime offers a simple urine test for some of the most common mycotoxins that cause illness, namely ochratoxins, aflatoxins, gliotoxin, and trichothecenes. Over the past five years, I have found this information invaluable for both diagnosing and orchestrating treatment for mold toxicity. It takes the guesswork out of it. Seeing a lab report that demonstrates the obvious presence of excessive mycotoxins, patients who have been ignored for many years finally feel vindicated. Even more important, this information helps us enormously in directing their successful treatment.

Because many mold-toxic patients have a *significantly compromised ability to detoxify,* I encourage them to do a "challenge" procedure when collecting urine specimens for the measurement of mycotoxins to improve our ability to detect mycotoxins in their urine. By "significantly compromised ability," I mean that the toxins have literally poisoned the very systems needed to remove these toxins—the liver, GI tract, kidneys, and lymphatic system—making it difficult for patients to mobilize those toxins (even though they are there) sufficiently to make the test positive. I find that using oral glutathione, 500 milligrams twice daily, for a week and *then* collecting the urine improves our diagnostic capabilities. So does using a procedure in which the patient sweats (sweating mobilizes toxins, too), such as sitting in an infrared sauna or a hot tub for ten to thirty minutes just prior to collection. Note that excessive sweating can indeed mobilize toxins and make patients feel worse, so they need to carefully monitor their bodies' responses and not overdo it. Some of my really sensitive patients can withstand the heat for only a few minutes, but it is still useful for getting better information from the lab.

I would like to add a cautionary note about the glutathione challenge. It is not unusual for toxic patients to mobilize toxins faster than their bodies can unload them, leading to an exacerbation of symptoms that may be intense. If this happens, the patient should stop taking glutathione and immediately send in the urine sample for analysis.

More recently, the Great Plains Laboratory began offering a urine mycotoxin test that appears to be more accurate and reproducible. That test is not exactly the same as the one offered by RealTime, so at the time of this writing, a combination of these two tests seems to give us the most complete picture of mycotoxins currently available. The methodologies of the tests from Great Plains and RealTime are different: RealTime uses ELISA technology, and Great Plains uses liquid chromatography and mass spectrometry to obtain data. As we continue to work with these laboratories, I hope to learn which tests will ultimately provide the best information.

After the urine sample is submitted for analysis, it takes only a short time to get the results. Elevated levels of any mycotoxin indicate a clear diagnosis and allow us to proceed with treatment immediately.

The use of urine mycotoxin testing has, for me, revolutionized our ability to diagnose and treat mold toxicity, adding a specificity that is vital to this work. Now that urine testing is available, the use of the blood tests for markers such as c4a, TGF-beta-1, and MMP-9 are much less useful because they are much less specific. Urine testing tells us exactly which mycotoxins are present and, as I will discuss next, how to treat them. Blood tests can suggest the presence of mold toxicity, but they do not allow us to focus on the component of illness that needs to be addressed first.

Treating Mold Toxicity

The Shoemaker Protocol

Dr. Shoemaker bases his treatment program on his diagnostic evaluation, which we have already reviewed. In its simplest form, this begins with evaluating the living environments for mold using the ERMI system and working on remediating the problem. (The ERMI system involves vacuuming dust from the room and sending that specimen to a lab, where the specimen is examined for thirty-six of the more common species of toxic molds. John Banta will review this process in more detail in Appendix A.) This may involve difficult family decisions because a mold-sensitive patient may be unable to live at home during treatment. Second, Dr. Shoemaker uses the binding resin cholestyramine (Questran) or its biochemical cousin, colesevelam hydrochloride (Welchol). These prescription medications are more traditionally used for the treatment of elevated cholesterol levels, but in mold-toxic patients, they bind to the toxin more strongly than the toxin binds to bile and thus pull the toxin out of the intestinal tract while the bile returns to the liver. This drastically decreases the load of toxin in the body and allows healing to begin.

Dr. Shoemaker recommends that his patients follow the low-amylose diet along with his full program. The low-amylose diet is a high-protein, low-carb diet in which the only carbohydrates allowed are corn products. He also includes the visual contrast sensitivity (VCS) test discussed earlier and repeats it at every visit to see how the treatments are progressing.

Let's go into Dr. Shoemaker's protocol in more detail.

The initial dosage of cholestyramine is one scoop (4 grams, which is the equivalent of 1⅔ teaspoons) four times a day. The timing matters, although it can be difficult to orchestrate. For optimal benefit, cholestyramine should be taken half an hour before a meal (with fat in the meal encouraged), and then the patient must wait another ninety minutes before taking any other supplement or medication, keeping in mind that cholestyramine is a strong binder and may interfere with the absorption of other materials if taken concurrently. Welchol works taken the same way, with the usual dosage being two capsules (625 milligrams each) three times a day.

Although cholestyramine is available by prescription from any pharmacy, the forms available from most pharmacies are not in line with the recommended treatment for mold toxicity. Questran comes in powdered form and contains a large amount of sugar. It is also available sweetened with Splenda or NutraSweet (which I do not encourage either). If cholestyramine is to be used, it should be obtained from a compounding pharmacy so that it is much purer and more in line with treatment. Giving sugar, as you might expect, would literally "feed" the mold and other microbes growing in the gut and therefore be contraindicated. Unfortunately, compounded cholestyramine can be expensive and is not always covered by insurance, adding to the cost of treatment.

Although cholestyramine has very few side effects, they include constipation, heartburn, and dyspepsia. Of these, constipation is the biggest concern; holding toxins in the intestinal tract for long periods clearly is not a good idea. Magnesium and vitamin C are the primary tools for working with this side effect. If a reaction to cholestyramine does occur, especially if it appears to be an exacerbation of the patient's current symptoms (patients often describe it as "like a Herx"), it indicates that the binder has pulled toxins into the patient's body faster than his or her compromised detoxification systems can clear them. As I emphasize throughout this book, this outcome is not helpful and must be carefully monitored. Any increase in toxicity will result in a worsening of the patient's condition, and the patient cannot push his or her way through an increase in toxicity successfully.

When I began this work under the tutelage of Dr. Shoemaker, I discovered that many of my patients seemed to be somewhat more complicated than those he saw. His patient population seemed to be more, if you will, "pure mold." Although I saw some of these kinds of patients initially, my patients tended to present with Lyme disease, Lyme co-infections, other biochemical perturbations, *and* mold toxicity. Many of these individuals were much more sensitive on every level to what they put in their bodies, and the majority could not take the doses of medication recommended by Dr. Shoemaker without being set back and made worse.

> **I cannot emphasize this enough: If some is good, more may not be better.**

For example, I have had very few patients who are able to take more than 1 or 2 teaspoons of cholestyramine a day. In fact, most of my patients must start with 1/16 or 1/8 teaspoon every other day in order to avoid provoking a setback. Once we determine a dosage that a patient can take comfortably, we increase it very slowly until we find a dosage that causes an exacerbation of symptoms, at which point we back off. I must stress that despite the best intentions to "fight through the side effects," this strategy does not work. Over time, the patient will simply become more toxic, and it may take weeks or months to get back to baseline. I cannot emphasize this enough: *If some is good, more may not be better.*

For those who are concerned about their inability to tolerate "normal" doses, it may be helpful to know that even minuscule doses of binders work well over time. I have had patients recover completely using just 1/16 teaspoon of cholestyramine every third day and half a tablet of chlorella (a type of freshwater algae commonly used as a binder for certain mycotoxins) every other day.

Often my patients question me: "Really, Neil, you think that pathetic dose is going to work?" And the answer is yes, it does, but they have to be patient with their bodies. An important concept here that many of my patients take months to grasp is: *Please listen to your body. It will clearly tell you how much it can handle.* If you ignore what your body is telling you, it will not go well. The flip side of sensitivity is responsiveness. The good news is that my experience has shown me that sensitive patients usually respond well to minuscule dosages as long as we are careful not to overdo it.

With cholestyramine as the mainstay of the binder component of his program, Dr. Shoemaker reports that many of his patients do great using just the binders and removing themselves from moldy environments. However, he discovered that commensurable bacteria growing in the sinus area, nicknamed MARCoNS (Multiple Antibiotic-Resistant Coagulase Negative Staphylococci), may prevent full recovery if not addressed at this stage. He cultures for these bacteria using a nasal swab and treats those patients who test positive with a combination of nasal sprays and antibiotics.

If patients do not respond as anticipated to his protocol, Dr. Shoemaker looks at their lab testing, especially c4a and TGF-beta-1, to see if they are continuing to demonstrate inflammation. If they are, he has a fairly elaborate series of additional interventions that he has found useful for those patients.

Finally, if patients still have not recovered as hoped, they are evaluated for the possible use of VIP nasal spray. While for many years it was thought that VIP would not work until the mold had been eliminated from the patient's home (and body) and the MARCoNS had been addressed, we are finding that for some patients, VIP sprays can be beneficial earlier in the course of illness.

While this protocol has been (and still is) instrumental in helping thousands of patients suffering with mold toxicity, I am increasingly finding it somewhat limited in scope and effectiveness. The pioneering work of Joseph Brewer, MD, which I will discuss next, has added multiple dimensions of treatment that have enabled me to help even more patients with improved results.

The Brewer Protocol (My Version of It)

Like Dr. Shoemaker, Dr. Brewer, an infectious disease specialist from Kansas City, begins with a careful, complete patient history and physical examination. Dr. Brewer then uses the test from RealTime Laboratories that evaluates families of the most common mycotoxins in the urine (see page 61). The test is easily done, but expensive. Urine is simply collected and mailed to the lab, and the turnaround time is fairly rapid—about one week. The new Great Plains test (also discussed on page 61) shows promise, but its turnaround time is a bit longer. If either of these tests is clearly positive, we have immediate verification of the diagnosis of mold toxicity and a clear starting point for which binders to use to begin the treatment process.

Dr. Brewer begins treatment with antifungal nasal sprays and the concomitant use of the appropriate binders, which I will describe in detail below. Carefully following the patient's clinical progress, he then adds oral antifungal treatment to eradicate what are believed to be the major areas of mold colonization: in the sinuses and gut.

This is a simplified overview, but the program turns out to be a lot more complicated to put into action, and I put my own spin on various aspects of both diagnosis and treatment. As I've said, my patients tend to be referrals from other physicians who have had difficulty managing these extremely sensitive individuals, so all of my comments are directed at that patient population. It is, if you will, the worst-case scenario. The very slow and deliberate protocol outlined here enables the maximum number of patients to respond without making them worse. Patients with stronger constitutions can handle more aggressive treatment and often progress faster, and that, of course, is ideal.

First, to reiterate, because I think this is so important: simply collecting a urine specimen may not be as accurate as we need it to be for diagnosis and treatment. Because they are toxic, many of the sickest patients have seriously compromised abilities to remove toxins from their bodies. The organs of detoxification—the liver, gut, kidneys, lymphatics, lungs, and skin—are often unable to keep up with the toxic load. This means that patients with severe symptoms are often *unable to release toxins into their urine* without a little help. When we first began urine testing, a small but significant percentage of my patients tested negative, even when they, and I, were certain that mold toxicity was present. Over time, we realized that if we did a sauna challenge (see page 61) and collected their urine thirty minutes after they got out, we would see elevated toxin levels in their urine, which more accurately reflected their diagnoses and helped us more accurately target their treatment.

As this process evolved, we found that we were still missing the diagnosis in some patients who tested negative even with the sauna challenge. We began the routine use of the glutathione challenge, also outlined on page 61. We found this approach to be more accurate, and considering the expense of the test, it is what I currently use to assist with diagnosis. Again, one caveat: my most sensitive patients often find that they cannot take that much glutathione without experiencing an exacerbation of symptoms. Glutathione does, indeed, facilitate the removal of toxins, but for many people, this means they are mobilizing toxins into their bodies faster than their bodies can release them, and they get worse. Although this helps confirm the diagnosis, it is not necessary to make a patient sicker. Once a patient starts to react badly, I stop giving glutathione and instruct the patient to send in urine for analysis right away.

Occasionally, the urine test does not confirm the diagnosis. In those cases, I have found it useful to repeat the test using a simple nasal washing and sending that material to a lab to be analyzed.

Once the lab processes the specimen, we get the results fairly quickly. It is important to understand what we learn from the report.

First, any elevated toxin level is clinically meaningful. It confirms the diagnosis and informs us how to start and direct treatment. The numbers detailing how much toxin is present may reflect a clear measure of the level of toxicity, but it is also a crude measure of the *patient's ability to detoxify,* so I have come to view the first test as the "tip of the iceberg." Those patients who have a compromised ability to detoxify, even with the provocation of a challenge test, may be able to excrete only a small amount of the toxin(s) that are present. When we repeat the test three to four months later, I must warn them that the numbers may be much higher, because, as they improve, their ability to detoxify improves simultaneously, and follow-up tests may more accurately reflect the actual toxin load (and types of toxins present). Following the mycotoxin results, which I find quite helpful, therefore requires understanding and interpretation as treatment unfolds, and information must be matched to the patient's condition as it evolves.

I have treated several dozen patients whose initial urine tests were negative but who have classic symptoms of mold toxicity and known exposure to mold at home or at work. When treated empirically (meaning, using my best medical judgment without having objective evidence to support that diagnosis), most of those patients recovered, and as they did, repeat testing showed large amounts of mycotoxins in their urine. I want to remind you that there are dozens of mycotoxins that we are not yet able to measure, so it may be necessary to treat some patients empirically at this time, based solely on a clear history and a review of symptoms.

FOLLOW-UP MYCOTOXIN TESTING

Urine mycotoxin testing allows us to accurately follow a patient's treatment. Numbers going up could mean improved detoxification, continued exposure to mold, or excessive treatment with binders or antifungal therapy. Numbers going down suggest that things are moving forward nicely. Because patients have varying abilities to prepare for testing with challenges (sweating, glutathione, or both), I encourage follow-up tests to be done under the exact same conditions so that we can compare apples to apples. Retested patients are usually being actively treated with binders and/or antifungal agents, so the question arises: Should the patient stop taking those before the next specimen is collected, and if so, for how long? Unfortunately, to my knowledge, this has not yet been studied, so we have no scientific answer at this time. RealTime recommends going off binders for a week prior to collection, and Great Plains recommends going off binders for forty-eight to seventy-two hours. I personally have not recommended going off binders at all during collection. I plan to do some research to measure urinary mycotoxins while continuing treatment, then measuring the same patient three days, then seven days off treatment, and I hope we can answer this important question soon.

Finding and Using the Best Binders

Carefully delving into the medical literature (which, like all literature searches, contains controversies), Dr. Brewer discovered information suggesting that certain toxins are bound, preferentially, by certain binders. To be more specific:

- **Ochratoxins** appear to be bound best by cholestyramine, Welchol, and activated charcoal.

- **Aflatoxins** are bound by activated charcoal and bentonite clay.

- **Trichothecenes** are bound by activated charcoal and possibly other binders. Chlorella and bentonite clay appear to be useful, but the medical literature is limited.

- **Gliotoxins** are bound by bentonite clay, *Saccharomyces boulardii* (a probiotic yeast), and N-acetyl cysteine (NAC).

The newest mycotoxin tests from the Great Plains Laboratory are now available. The names of these toxins are difficult to pronounce and related to the toxins already mentioned, so I will try to simplify it all here. They include sterigmatocystin, which is related to aflatoxins and thus treatable with the same binders, and roridin E and verrucarin A, which are macrocyclic trichothecenes and thus bound well by the binders used for other trichothecenes. In addition, we are now able to measure zearalenone, which is made by *Fusarium* fungi. Very little research is available about the most effective binders for some of these newly available mycotoxins, but from the clinical response we are seeing, cholestyramine, chlorella, activated charcoal, and bentonite clay appear to be effective.

This information allows us to be more specific in prescribing binders to help the body remove these toxins. Using the the lab report as a guide, I usually start with the binder(s) that will bind to the toxin present in the largest quantity in the urine. Be aware that this target may shift with follow-up urine tests. For example, say a patient has an ochratoxin level of 4.85 (normal being anything below 1.8), an undetectable aflatoxin level, a trichothecene level of 0.95 (normal being above 0.18), and a gliotoxin level of 5.62 (normal being above 0.5). A quick review of these numbers suggests an excess of ochratoxins of about two-and-a-half-fold, trichothecenes four-fold, and gliotoxin eighteen-fold. (This is a fairly typical report.) Logic, therefore, would be to start treatment with a focus on gliotoxin, incorporating components for trichothecenes and ochratoxins in that order, if possible.

Let me use my own experience with mold and my own lab results as an example of how to read these reports and proceed with treatment.

NEIL'S STORY

Two summers ago, my wife and I noticed mold growth on the ceiling of an outbuilding that we use as a workshop. Because I had been exposed to mold off and on over the years with no adverse reactions, I assumed (incorrectly) that I was one of the lucky folks who can be around mold and not be affected by it. I proceeded to scrape it off and use bleach and Kilz paint to cover it. In retrospect, I probably did everything wrong.

There were no immediate, obvious difficulties, but some time after I finished, I noted the onset of persistent loose stools and diarrhea, with occasional mild abdominal cramping. This was similar to what I experienced when I developed a gluten intolerance about twenty years ago, so my first assumption was that I had developed another food allergy. I systematically removed virtually everything I ate with any regularity from my diet and noticed no improvement. The loose stools continued.

I consulted with many expert functional medical doctors and took all of their advice. I tried every supplement, enzyme, and probiotic anyone could think of and tested my stools using excellent laboratories (which revealed nothing). As this persisted into the late fall, with no other symptoms, I sought consultation with a gastroenterologist, who did a colonoscopy that showed "mild" colitis. Testing of my stool at that time revealed the presence of a parasite called Blastocystis hominis. *This is a curious microbe in that some people can harbor it and exhibit no ill effects whatsoever, while others experience a wide array of symptoms, like mine. So I took several rounds of increasingly intense antiparasitic medications, with no benefit whatsoever. While my symptoms were annoying, they were only mildly debilitating. As they persisted, however, the mystery of their cause continued to bother me.*

By late spring, perhaps ten months after all this started, my wife recalled that my symptoms came on fairly soon after I cleaned up the mold. I believe that God has both a sense of humor and a sense of irony, and once this possibility came to light, I underwent urine mycotoxin testing and was clearly positive in several categories. I immediately began treatment with the specific binders for my toxins, and within six weeks my symptoms were gone. Relieved that I had taken care of the source of my problem using binders alone, I continued taking them for many months. My mycotoxin levels decreased but did not completely resolve (although my symptoms were still gone), and it became clear that to get well, I needed to go on antifungal nasal sprays and oral treatment to accompany the binders. As of this writing, I am on both and am scheduled to retest shortly. I did a mycotoxin urine test using provocation from both the Great Plains Laboratory and RealTime Laboratories, and here are the results:

Code	Test	Specimen	Value	Result	Not Present if less than	Equivocal if between	Present if greater or equal
E8501	Ochratoxin A	Urine	1.97600 ppb	Equivocal	1.8 ppb	1.8 - 2.0 ppb	2.0 ppb
E8502	Aflatoxin Group (B1, B2, G1, G2)	Urine	2.78000 ppb	Present	0.8 ppb	0.8 - 1.0 ppb	1.0 ppb
E8503	Trichothecene Group (Macrocyclic)	Urine	0.11100 ppb	Present	0.02 ppb	0.02 - 0.03 ppb	0.03 ppb
E8510	Gliotoxin Derivative	Urine	3.44400 ppb	Present	0.5 ppb	0.5 - 1.0 ppb	1.0 ppb

Comment: Due to the commercial unavailability of previously used standards, RealTime Lab has validated a new, more sensitive standard for Trichothecene Testing. Effective 11/13/2017, all results are reported using the new standard. Please note the new values for cutoff levels.

RealTime Laboratories, Inc.

Mycotoxin Panel Report

Great Plains Laboratory, Inc.

GPL-Mycotox test

MycoTox Profile

Metabolite	Results (ng/g creatinine)	Common Range of Positive Results	
Aflatoxin M1	0	1.3 - 3	
		↑1.3	3↑

Metabolite	Results (ng/g creatinine)	Common Range of Positive Results	
Ochratoxin A	2.13	1.2 - 5	
		↑1.2	5↑

Sterigmatocystin	0	0.1 - 2.25	
		↑0.1	2.25↑

Metabolite	Results (ng/g creatinine)	Common Range of Positive Results	
Zearalenone	0	0.5 - 2	
		↑0.5	2↑

Roridin E	0	0.75 - 2.25	
		↑0.75	2.25↑

Metabolite	Results (ng/g creatinine)	Common Range of Positive Results	
Verrucarin A	0	0.5 - 1.2	
		↑0.5	1.2↑

Enniatin B	0	0.09 - 0.45	
		↑0.09	0.45↑

From Great Plains, you can see a level of zero for most of the mycotoxins tested, but ochratoxin was 2.13 (positive is 1.2 or higher). The RealTime test shows a mildly elevated level of ochratoxin A at 1.97 (anything above 1.8 is positive), which agrees with the Great Plains test, but also an elevated aflatoxin of 2.78 (the upper limit of normal is 0.8). Trichothecene was 0.11 (the upper limit of normal is 0.02), and gliotoxin was 3.44 (the upper limit of normal is 0.5).

What can I learn from this? First, although my symptoms are gone, I am still mold-toxic. From years of experience with patients, I have learned that ignoring this toxicity would likely lead to a regrowth of mold and Candida *inside me and eventually make me sick again. So I am taking these results seriously. Second, you can see that both labs measure different kinds of toxins, so other than ochratoxin A (which I have found quite consistent from patient-to-patient measurements when both labs were used for the same specimen), the other toxins measured are not the same. Third, each lab measures several different toxins. RealTime measures gliotoxin, and Great Plains measures enniatin B and zearalenone. What this means is that working with both laboratories, which utilize different methods of analysis, gives us a more comprehensive picture of what is going on inside.*

The next section goes into more detail about treatment, and you will see how the knowledge of which mycotoxins are present informs us how to proceed with treatment. Unless we can maximize the information available to us and treat all of the mycotoxins that are present, treatment may not work.

For reasons I will discuss shortly, I start with the gentlest binder, so I begin most patients with one *Saccharomyces boulardii* capsule, 8 to 10 billion units, with one meal per day. For more sensitive patients, I start with a lower dose of 3 to 5 billion units per capsule and try one-quarter capsule at first, then slowly increase the dosage if it is well tolerated. After a few days, if no negative reaction has occurred, the dosage can slowly be increased to one capsule with each of two meals a day, and finally to one capsule with each meal. Most of my sensitive patients can take *Saccharomyces* without difficulty, but as noted previously, we proceed with caution.

Next, I add bentonite clay. Most patients start with one capsule containing about 500 milligrams. Here is where we might start noticing problems in sensitive patients. Many of my patients react strongly to even a single capsule of clay with an intense exacerbation of their usual symptoms: an increase in fatigue, anxiety, cognitive impairment, joint pain, and so on. If this happens, I instruct them to stop taking the clay immediately, wait until the exacerbation has completely resolved, and then try again, taking either a half or a whole capsule every second or third day. If the patient can handle this dosage, fine. If not, we look for a liquid preparation. I have found that the Yerba Prima Great Plains Liquid Bentonite Clay can be started at 1⁄16 or 1⁄8 teaspoon taken once daily and slowly worked up to 1 teaspoon, or the *maximum dose the patient can take with complete comfort.*

It is critical for patients to understand that they should not tolerate any worsening of symptoms: worsening means that they are pulling more toxin into their bodies than their bodies can process, and over time they will have an unacceptable exacerbation. Most of my patients have been high-functioning, active, intelligent professionals, and it is hard for them to grasp this point. Often they tell me, "Neil, I'm just a little bit worse, so I will tough my way through it and get better faster." Oh no they won't. Hundreds have tried this with no success.

I cannot emphasize enough that in treatment, *if some is good, more is not necessarily better.* Patients must honor their ability (or lack thereof) to detoxify or they will get worse. Over time, as the toxins are excreted, the patients will be able to do more—lots more—and that is a barometer of improvement.

One of the most confusing aspects of treatment for both physicians and patients is the exacerbation of symptoms, sometimes profound, with the use of binders alone. I would like to make an effort to describe what I think is happening here.

The difficulty is that the word *binder,* though accurate, has the connotation that the toxin is being strongly, completely "stuck" to the binder being used, such as activated charcoal or chlorella. That's not exactly correct. Take charcoal, for example. Charcoal *adsorbs* toxins to its surface. Adsorbing is not exactly binding: it reflects being loosely connected, kind of like static cling. Binding, therefore, is not a one-time gripping-like-a-vise-and-won't-let-go process, but an active, fluid relationship.

Another analogy, recently given to me by one of my patients, is that binding is like a magnet attached to a pile of iron filings. As the magnet drags the filings along, particles drop off with regularity. This means that binders have the ability to pull toxin into the bloodstream at a faster rate than the body's detoxification system can clear it, which means that the *patient is more toxic,* at least temporarily. This is not a rare event; it happens with almost every patient, repeatedly, unless he or she recognizes what is happening and reduces the level of binder to the point that no worsening whatsoever occurs.

The bottom line, which is of the utmost importance in treatment, is that both patient and physician must monitor the patient's process and constantly tweak it. This requires much more supervision and input than many patients (and physicians) are used to. It is not simple or easy, although it might sound that way. After all, how hard can it be to take chlorella? It says right there, on the bottle, to take fifteen tablets twice a day. The vast majority of my patients, however, can take only three or fewer tablets once a day; many can take only a portion of one tablet every other day. Keep in mind that my patients tend to be more sensitive and compromised than most, and what I have written is chiefly for their benefit. Patients who have heartier constitutions and who have not been ill for as long may be able to take much more, more easily. *Treatment must be individualized.*

With this overview, I would like to provide dosage ranges for the various binders we use, for the very sensitive to the much more tolerant. Note the huge ranges.

BINDER	DOSAGE	TIMING	NOTES
Cholestyramine	$\frac{1}{16}$ teaspoon of powder once every other day to 4 scoops daily	Taken half an hour before a meal; the patient must wait another 90 minutes before taking anything else (such as a medication or supplement to which it might bind, decreasing its availability to the body)	
Welchol	$\frac{1}{4}$ tablet once daily to 2 tablets (625 milligrams each) 3 times a day	Taken half an hour before a meal; the patient must wait another 90 minutes before taking anything else (such as a medication or supplement to which it might bind, decreasing its availability to the body)	
Activated charcoal	A portion of a capsule (500 milligrams) once daily to 3 capsules a day	Best taken on an empty stomach, 2 hours before or after eating	
Bentonite clay	$\frac{1}{16}$ teaspoon of liquid to 3 capsules once daily	Best taken on an empty stomach, 2 hours before or after eating	
Chlorella	$\frac{1}{8}$ tablet every other day to 3 tablets (200 milligrams each) once daily	Best taken on an empty stomach, 2 hours before or after eating	Use "glass-grown" chlorella; "natural" chlorella is often grown outdoors in ponds and, as a binder of toxin, may contain excessive amounts of arsenic and aluminum.
Saccharomyces boulardii	$\frac{1}{4}$ capsule (3 to 10 billion units each) with one meal a day to 1 capsule with each meal	Best taken with food	
OptiFiber Lean	$\frac{1}{4}$ scoop mixed in water once daily to 1 scoop 3 times a day	Best taken on an empty stomach, 2 hours before or after eating	This binder, which appears to be useful as a generally applicable binder for all of the named mycotoxins, has been well tolerated by my more sensitive patients. I am grateful to Jill Carnahan, MD, for this valuable suggestion.

To give you a realistic idea of how this process works, I usually spend six to eight weeks figuring out the proper dosage of binders for each patient before I proceed to the next step. It can take longer. There is no way to rush this process, and, at the risk of belaboring this point, each patient must move forward at his or her own pace. To ignore this guidance is to risk a serious setback that can last for weeks. Remember, if some is good, more is not necessarily better.

Antifungal Treatment with the Brewer Protocol (with My Modifications)

With the binders up and running, the body is better protected from the sudden release of toxins that may occur when substances that can kill mold are taken. Most of the antifungals that we use work by punching holes in the wall of the mold cell, killing the mold but, in the process, also releasing the contents of that cell, which contains mycotoxins. This is frequently referred to as a "die-off" reaction. This makes the patient vulnerable to the sudden influx of mycotoxins and can (and often does) cause an exacerbation of symptoms identical to those that can be triggered by the use of too much binder. This, too, needs to be respected, and the caveat here is to go very slowly and gently and make sure that the patient remains comfortable throughout the process.

Dr. Brewer begins with an antifungal nasal treatment program. This, too, is individualized. When Dr. Brewer and I started these treatment protocols, we used nasal sprays of amphotericin B, which researchers at the Mayo Clinic had used in their groundbreaking research in 1999. We discovered that although amphotericin B worked, and worked well, for many patients it caused either too intense an inflammatory reaction (making it hard for them to breathe through their congested sinuses or causing pain or nosebleeds) or killed the mold faster than the patients could process the toxins that were released as a result. About 60 percent of patients can handle amphotericin B well; the rest cannot. We also discovered that if we combined the nasal spray of amphotericin B with a nasal spray of EDTA, which is an excellent material used to dissolve biofilm (see the next page), congestion was magnified and made treatment too difficult for many patients to tolerate.

It turns out that how patients handle the binders is a good guide to how they will handle the nasal and/or oral treatments. Those with stronger constitutions who can take all the binders we give them at full dosages usually do well with an amphotericin B nasal spray. Those who can barely handle even minute amounts of binder require the gentlest treatment: nystatin nasal spray. This applies almost universally to my sensitive patients, for whom the use of nystatin nasal spray is our go-to first treatment. Those patients who fall somewhere in between often do well with a 2 percent ketoconazole nasal spray or a 1 percent itraconazole nasal spray. Individuals vary, of course, and we have to carefully monitor their responses to ensure that the treatment is being handled comfortably. No worsening or exacerbation of symptoms is acceptable, as toxins will accumulate and the patients will get worse if we ignore such a response. The longer they put up with any degree of worsening, the longer it will take for their bodies to right themselves when we stop the irritating treatment.

I have found that starting with hydrosol silver (often referred to incorrectly as colloidal silver) as a nasal spray before using the antifungals allows me to get advance notice of sensitivity. I recommend one spray of Argentyn 23 in each nostril. As noted, very rarely patients get minor irritation from that material and, also rarely, a "die-off" requiring us to back off, then decrease the dosage to once every two or three days. The vast majority of my patients—even the most sensitive ones—do well with it.

Then I begin the antifungal nasal spray, choosing one that is likely to be tolerated based on the patient's experience with binders. It works best if hydrosol silver and the antifungal are taken at the same time, as they are synergistic. Once this treatment is tolerated, I add a nasal spray that will help dissolve biofilm. Colonized mold in both the sinus and the gut tries to protect itself from the immune system and from treatment by secreting a thick layer of biofilm (kind of an amorphous glop) around itself, and we have learned that we need to address this layer in order to make progress. The two major materials used here, made by compounding pharmacies (as are the nasal sprays), are a chelation spray that contains several materials, primarily EDTA, and a BEG (bacitracin/EDTA/gentamicin) spray, which also contains EDTA. Treatment begins with one administration daily. The BEG spray can be a bit gentler for sensitive patients; we find that the chelation spray causes an unacceptable amount of nasal congestion in some people.

More recently, I have been leaving the gentamicin out of the BEG spray, as it can cause side effects and may, in the long run, cause more sinus bacterial resistance. At the 2018 American Academy of Environmental Medicine meeting, Michael Gray, MD, one of the pioneers of mold toxicity treatment, shared that he has found that 0.06

percent amphotericin B is well tolerated in both the sinus and gut areas and works well. I have been utilizing his suggestion with many of my patients and find it valuable. At that same meeting, Irene Grant, MD, another pioneer in this area, shared that very few treatments work on the hyphae forms of mold except amphotericin B, giving additional weight to the use of that material as a main ingredient of treatment if the patient can handle it.

Some patients can tolerate the nasal sprays only every other day or every third day, but others can handle them twice or even three times a day. Every step of the treatment program must be individualized based on each patient's response. (I know I keep reiterating this, but I cannot overstate its importance.)

Once the nasal sprays are up and running, we turn to oral treatments to address mold that is colonizing the intestinal tract. For this task, we use the same kind of materials that we use for the sinuses: we start with hydrosol silver in the form of Argentyn 23, ½ to 1 teaspoon twice daily, then add an antifungal medication and a biofilm-dissolving agent. Here, Dr. Brewer has primarily used itraconazole (Sporanox), based as always on patient tolerance. With my very sensitive patients, I start with 100 milligrams every two weeks and slowly, as tolerated, work it up to once a day, then to twice a day or even two doses twice a day if the patient is stronger. If the patient has the clear presence of gliotoxin, which is made by both *Aspergillus* and *Candida,* I add 500 to 600 milligrams of NAC once or twice daily and nystatin tablets (500,000 u each) two to four times daily. For sensitive patients, I start with one-quarter to one-half of a nystatin tablet and increase the dose slowly as tolerated. I also add one of several agents that help dissolve biofilm. My two favorites are InterFase Plus from Klaire Labs, taken one to three times a day, and MC-BFM from Beyond Balance, taken twice daily. There are quite a few additional treatments that are not standard components of the Brewer Protocol but that I find quite helpful in speeding up recovery, which I will get into next. Dr. Brewer also recommends the use of a nasal spray of NAC to help break up biofilm and speed the recovery from gliotoxin.

There is a lot of controversy about whether or not gliotoxin is made by *Candida.* The medical literature is not helpful here because quite a bit of supporting evidence can be found for either side of the question. My explanation for this discrepancy is as follows: it is not easy for molds (or *Candida*) to make toxins; the manufacture of toxins requires a significant output of cellular energy. Generally, molds will not make toxins unless they feel threatened. In the natural world, this threat would come primarily from other molds. In our bodies, this threat comes from taking antifungal materials that are designed to kill the molds. Therefore, in a laboratory setting, growing *Candida* on nutritionally rich plates, at a perfect temperature, in the dark, does not stimulate it to produce toxins. In the body, where *Candida* is threatened, it does. I confess that this is a speculative hypothesis, but it has a certain logic and is supported by some science. The bottom line is that both Dr. Brewer and I have found that including treatment for *Candida* early in the antifungal part of this program appears to improve treatment overall.

STEP ONE Evaluate your environment (home, work, car) for mold. While there are many methods for testing for mold, a relatively cheap and accurate option is the ERMI test.

STEP TWO Do the mycotoxin test through either Great Plains Laboratory or RealTime Laboratory to see if you have mycotoxin issues. To improve diagnostic capabilities, use oral glutathione, 500 milligrams twice daily, for a week prior to taking the test and sweat for 10 to 30 minutes with a sauna or hot bath 30 minutes before collecting your specimen.

STEP THREE Add a binder specific for the mycotoxins found through your testing. See the binders chart on page 73 for specifics.

STEP FOUR Add oral and/or nasal antifungals to treat mold that may have colonized in your body.

It can take a year or more for this treatment program to work, so the next part of my job is to keep the patient motivated and feeling positive about the results. Dr. Brewer has noted that his patients who have taken the nasal sprays the most consistently have done the best clinically. He has published two papers, one demonstrating the effectiveness of amphotericin B nasal spray and the other the effectiveness of nystatin nasal spray, resulting in improvements for 94 and 89 percent of his patients, respectively.

A word of caution: Many, many "binder" treatments have been proposed but have not been shown to be specific for mold toxins. That does not necessarily mean that they don't work, just that there is little research to confirm that they will work *for mold.* Many practitioners utilize a lot of their favorite "toxin binders" without realizing that those binders may be great for some toxins, but not for mold. This is of great concern to me because I am seeing more and more referrals from healthcare providers who have been utilizing binders that might be great for heavy metals, but not necessarily for mold, and are unaware that there may be no crossover in effectiveness. Our knowledge, however, continues to grow. One of my patients recently sent me an article by T. S. Kahlon, M. H. Chapman, and G. E. Smith, who tested the binding of bile acids (which carry mold bound to them) by okra, beets, asparagus, eggplant, turnips, green beans, carrots, and cauliflower and found okra to be the best food at assisting in the binding of these toxins.

Some researchers in the mold toxicity field are not convinced that mold can colonize inside the body and make toxins on an ongoing basis. Additional research will clarify this point. The bottom line for me (and for Dr. Brewer) is that over time, the majority of our patients who had not responded to other kinds of treatment have responded well to this program and have seen their health restored.

I find that in those patients who do not respond to treatment as quickly as we would like, the primary problem is that they are still being exposed to mold toxins. It is necessary for these patients to dig deeper into their work or home environments to find out where the mold is. Until the exposures are brought to light, improvements in health will not occur.

HEATHER'S STORY

Heather was a twenty-year-old woman who was referred to me by the Cleveland Clinic because she had developed symptoms that were unresponsive to treatment. She had, over several years, developed extreme fatigue accompanied by fluctuating joint pains, cognitive dysfunction, dizziness, paresthesias, and anxiety. But the most unusual and troubling symptom was that she would, without warning, suddenly collapse and fall over. She described this as "my whole body gets paralyzed." These debilitating episodes were occurring twenty to thirty times a day and required the presence of a family member to watch with vigilance, as Heather was in imminent danger of injuring herself with each episode. She had also been diagnosed with postural orthostatic tachycardia syndrome (POTS) at Johns Hopkins. Multiple neurologists had examined her, and, despite the intensity of these events, her EEG tracings were normal, and no clear diagnosis was offered.

When Heather arrived in my office for her first exam in September 2015, I witnessed firsthand several of these episodes, and they were indeed alarming. All of a sudden, in the middle of a sentence, she simply would keel over to one side. Although I consider myself to have excellent reflexes, I never could have caught her; fortunately, Heather's mother was carefully positioned next to her to prevent falls. Heather emerged from these episodes quickly, and we were able to resume our discussion without missing a beat. This episode was not a seizure—it was more like a syncopal (fainting) episode, but one that appeared to be more neurologically initiated.

I thought that the possibility of mold toxicity best explained the totality of Heather's clinical picture. While she did have a borderline positive Lyme test, her urine mycotoxin test for trichothecenes came back at 0.30 (the upper end of normal being 0.18). We began treatment with chlorella and quickly realized how sensitive she was. She could barely tolerate one-quarter of a 200-milligram chlorella tablet. Any attempt to increase the dose resulted in an obvious exacerbation of these fainting episodes, along with an increase in fatigue, anxiety, and cognitive impairment. Very, very slowly, she was able to increase her chlorella dose so that by December she was able to tolerate a whole tablet. On even that tiny dose, her syncopal episodes had decreased to only a few per day, and she was able to make greater strides in her ability to tolerate treatment. By January she could handle two chlorella tablets a day, and her episodes were down to one or two per day. She soon was able to tolerate Argentyn 23 nasal spray, nystatin nasal spray, and BEG nasal spray. By March 1, her episodes had *decreased to only one or two per week, and her energy and cognition had improved to the point where she could resume online college classes.*

By the end of March, repeat urine mycotoxin testing showed an increase in ochratoxins to 1.91 (anything above 1.80 is considered toxic; ochratoxins were not seen in her initial test results), a rise in trichothecenes to 0.65, and, most notably, a gliotoxin level of 7.76 (the upper limit of normal is 0.50). Keeping in mind that Heather was much, much better by this time, this increase in mycotoxin levels is the rule rather than the exception. I interpreted it to mean that Heather's ability to eliminate toxins had improved to the point that she could now excrete these toxins in her urine. These new results probably provided a more accurate reflection of the toxic load she originally had, but her inability to excrete toxins at that time limited her initial results. The much higher numbers also would correlate well with the severity of her illness when she first presented to me.

Heather's case is rather typical of both mold presentation and treatment. Her symptoms could be explained as unusual neurological manifestations of mold toxicity and have resolved completely over the ensuing three years. Her initial marked sensitivity allowed the use of only tiny doses of binder, yet that treatment resulted in fairly rapid improvement and ultimately the ability to tolerate a more complete treatment program. Treatment resulted in virtually complete resolution of all symptoms, and Heather was able to return to college full-time and successfully complete her courses. Healing was slowed on two occasions by re-exposure to mold while at college, which was dealt with promptly.

Detoxification Strategies

For some patients, detoxification might actually be Step 1 in the treatment process. Many patients come to me having been sick for so long without being diagnosed correctly that they are extremely toxic and sensitive. Some cannot begin any form of treatment until we address their ability to detoxify. I often start these patients on a group of supplements and other treatments designed to accomplish this task. While there are many such supplements on the market, I am going to share the ones that I have found to be the most effective.

I begin with a product from Beyond Balance called Tox-Ease GL, which helps specifically with the liver and GI tract, using one or two drops once or twice daily and working up slowly, as tolerated by the patient, to five to ten drops twice daily. Then I add two homeopathic materials from BioResource, called Itires and Renelix, again starting with one or two drops twice daily and working up to five to ten drops twice daily as tolerated.

One of my favorite treatments, especially for sensitive patients, is the use of intravenous phosphatidylcholine (discussed in more detail in Chapter 15). Really sensitive patients usually cannot tolerate the typical doses, which are 10 to 15 cc (two to three vials) intravenously. I start my ultrasensitive patients on 0.5 to 1.0 cc and then slowly work up to higher doses as tolerated. At the risk of sounding like a broken record (my age is showing with this analogy), even tiny doses of phosphatidylcholine run the risk of provoking an exacerbation unless we proceed with caution—as is the case with every other treatment I describe.

I have also found that the use of ozone can be beneficial. Ozone can be delivered in a variety of forms, including nasal insufflation, ear insufflation, rectal insufflation, and intravenously. The most useful forms for my patients have been nasal and ear insufflation.

Another promising treatment, developed by Ty Vincent, MD, is called low dose immunotherapy (LDI). A variety of materials have been used to decrease the inflammatory response and improve the function of the immune response to mold toxins and *Bartonella*, among many other possible insults to the body. I discuss these materials in more detail in Chapter 10.

Another advance in the treatment of the immune system is the recent availability of Transfer Factor Enviro from Researched Nutritionals, designed to enhance the body's ability to process *Aspergillus, Penicillium, Fusarium, Cladosporium,* and *Candida.*

Each of these treatments should be instituted slowly, one at a time, to carefully assess whether it is being well tolerated. Once a small dosage is seen to be well tolerated, the patient can move on to the next treatment.

Additional treatments that are often valuable at this stage include Epsom salt baths and infrared saunas (two or three times a week, carefully monitoring the duration of exposure; sensitive patients may be able to handle only ten to fifteen minutes at first, and I don't advise going over thirty minutes at a time). Oil pulling, using an oil with a good "mouthfeel" like organic coconut oil or olive oil, can be helpful, too, starting at three to five minutes once a day and slowly working up to twenty minutes a day if possible. The use of the Beyond Balance product Mycoregen and the Byron White Formulas

> **Many patients come to me having been sick for so long without being diagnosed correctly that they are extremely toxic and sensitive. Some cannot begin any form of treatment until we address their ability to detoxify.**

A-FNG also can help patients begin the process of improving immune response. (See Chapter 10, which covers the immune system in more detail.)

The role of methylation, which is profoundly affected by mold toxins, in detoxification is discussed in Chapter 14.

Summary

I have shared with you my observation that mold toxicity is far and away the most underappreciated cause of sensitivity and toxicity in my patients. It is not rare; it is estimated that well over a million people in the United States may be affected by mold toxicity in some way. We are just beginning to understand just how massive a problem it is and to develop appropriate diagnostic and treatment strategies. This chapter was not intended to be the definitive answer, but to provide the best starting point for addressing this issue that is currently available.

My patients who have become sensitive and toxic are painfully aware of how their descriptions have been received by medical professionals and by friends and family members. To those who have not worked with these patients, it seems impossible that anyone could react so badly to minuscule doses of supplements and medications. But they do.

The starting point for diagnosis, considering the adage "common things occur commonly," would be to look closely at the possibility of mold toxicity in all of these unfortunate individuals. It goes without saying (but I am going to say it anyway) that if you don't look for it, you are unlikely to find it. What is really important is that mold toxicity is treatable, so please do not overlook it.

CHAPTER 4

Bartonella Infection
Sounds Like an Italian Liqueur, but…

Bartonella is a unique bacteria that has a nasty tendency to insinuate itself inside cells (or to "go intracellular"), which can cause a variety of infectious symptoms for the unfortunate individual who is wrestling with it. *Bartonella* infection is sometimes called *bartonellosis,* and it is often referred to by the names of two better-known infections: cat scratch fever, caused by *Bartonella henselae,* and trench fever, caused by *Bartonella quintana.* Lyme specialists often use the term BLO (*Bartonella*-like organism) to emphasize the fact that there are so many known *Bartonella* species that can cause illness (most of which cannot be tested for clinically) that the specific *Bartonella* organism that is causing a patient's illness may be impossible to identify with precision. This inherent lack of precision, as you might expect, prevents some physicians from embracing this diagnosis, as it is "too vague." Our difficulty in putting a precise label on this infection in no way changes that fact that it exists and that we need to treat it.

Getting down to basics: *Bartonella* is thought of as an *opportunistic infection.* What that means is that if the immune system is working well, the person harboring this bacteria can deal with it easily, and it will not cause any trouble. However, if the immune system is weakened for any of several reasons—such as mold toxicity, Lyme disease, a surgical procedure, childbirth, a severe flu, or an intensely stressful life event—the immune system loses its ability to keep *Bartonella* in check, and it starts to grow. Once this happens, *Bartonella* has unique properties that can distract the immune system in a variety of ways so that it can grow and thrive inside the body without being subject to the normal attacks of an intact immune system. One of the more obvious ways is that it can leave the circulation and "burrow" into cells so that the immune system (and the antibiotics and herbal remedies that might be used to treat it) has much less of an effect on it.

How can a person acquire *Bartonella*? About 40 percent of cats carry this bacteria, so a cat scratch or bite can allow this microbe to enter the body, where it will lie dormant until the immune system weakens and it can begin to thrive. *Bartonella* also can come from the bites of ticks, fleas, sand flies, and mosquitoes.

With the CDC formally recognizing, in 2013, the epidemic nature of Lyme disease, noting that 300,000 new cases are discovered every year, the presence of Lyme has become the most common explanation for how a patient has acquired *Bartonella.* After a tick bites a person, takes a "blood meal" into its stomach, and lets it sit there for a while, it injects the contents of its stomach back into that individual. (Some doctors believe that the meal sits in the tick's stomach for at least twelve to twenty-four hours, and that if you remove the tick quickly after being bitten, you will be "safe." The tick, however, does not have a wristwatch and sometimes will release its contents much sooner, so there is no time frame that is truly safe.) There have been many studies of the contents of what a tick injects. Joseph Burrascano, MD, a recognized expert in the diagnosis and treatment of Lyme disease, calls it "nature's dirty needle," which is an apt description. In addition to *Borrelia burgdorferi* (the scientific name of the Lyme microbe), a tick can inject *Bartonella,* the parasite *Babesia,* the microbes *Ehrlichia, Anaplasma,* and a variety of *Rickettsia* (such as Rocky Mountain spotted fever), and other infective agents, including viruses. Nasty, huh?

The infective agents within ticks' stomachs have been studied by many laboratories and vary a bit from region to region. Here in northern California, where I live and practice medicine, 44 percent of ticks carry *Bartonella,* 40 percent carry *Borrelia,* 10 percent carry *Ehrlichia,* and 8 percent carry *Babesia.* So you can see that it is easy to acquire several infections from one tick bite, hence the term *co-infections.*

The subject of Lyme disease with its co-infections is a huge one, far beyond the scope of this book. The reason I am focusing on the *Bartonella* microbe is that it appears to be uniquely related to the development of toxicity and sensitivity in many of the patients who contract it. When a patient presents with intense sensitivity and reactivity, *Bartonella* infection is near the top of my diagnostic list of things to explore.

Symptoms of Bartonella Infection

As with many chronic infections, the general symptoms of fatigue, malaise, swollen lymph nodes, sore throat, and brain fog are present in *Bartonella* infection, sometimes to an intense degree. Though important, these symptoms do not prompt us to look specifically in the direction of *Bartonella*. More unique to *Bartonella* infection are the following (described in more detail on the next page):

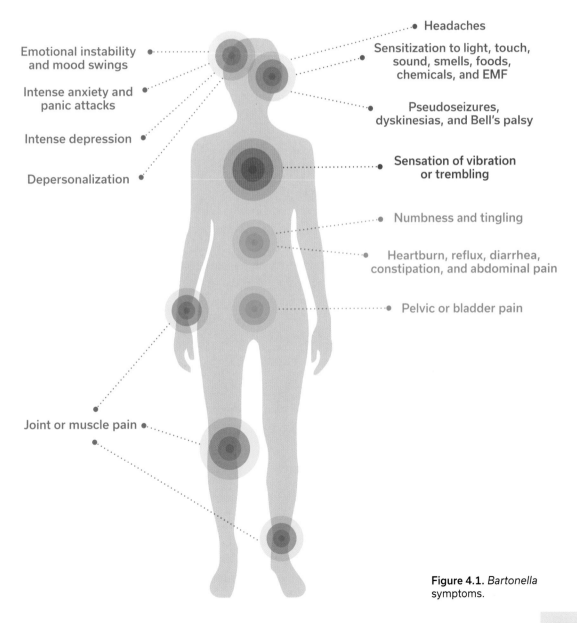

Emotional instability and mood swings

Intense anxiety and panic attacks

Intense depression

Depersonalization

Joint or muscle pain

Headaches

Sensitization to light, touch, sound, smells, foods, chemicals, and EMF

Pseudoseizures, dyskinesias, and Bell's palsy

Sensation of vibration or trembling

Numbness and tingling

Heartburn, reflux, diarrhea, constipation, and abdominal pain

Pelvic or bladder pain

Figure 4.1. *Bartonella* symptoms.

- Intense emotional instability and mood swings

- Intense anxiety, including "out of the blue" panic attacks

- Intense depression with occasional feelings of hopelessness or despair

- Depersonalization: a sense that one does not feel like oneself or is not present in one's own body

- An uncomfortable sensation of vibration or trembling, not visible to the naked eye, often residing in some part of the spine. While patients frequently describe this sensation as "weird," it is fairly typical of *Bartonella* (as well as mold toxicity).

- A perception of numbness and tingling, referred to medically as *paresthesias,* especially in areas of the body not typically associated with the peripheral nerves. To be more specific, neurologists often see paresthesias in the hands or feet but are baffled if they are felt in the face, abdomen, or mid-back areas. "Atypical" areas of paresthesias would immediately lead us to think of *Bartonella* (or mold toxicity) as a possible cause.

- A wide variety of neurological disorders, including pseudo-seizures (body movements that look like seizures, but without the abnormal electrical activity in the brain that is characteristic of seizures), dyskinesias (unusual twisting, spasmodic movements), and Bell's palsy. I would like to point out that a neurologist's diagnosis of "atypical" multiple sclerosis (MS), Parkinson's disease, Alzheimer's disease, or amyotrophic lateral sclerosis (ALS) is a tip-off to look for *Bartonella* (or mold toxicity).

- Sensitization to light, touch, sound, smells, foods, chemicals, and electromagnetic fields (EMF)

- Headaches localized to the back of the head (the occipital region)

- Joint or muscle pain that moves around rather than staying in one area (often a description of fibromyalgia)

- Gastrointestinal symptoms such as heartburn, reflux, diarrhea, constipation, and abdominal pain

- Pelvic or bladder pain

Patients describe their symptoms with intense, vivid language and gestures that are not typical of other infections. For example, while pointing to their heads with both hands and making facial expressions indicative of intense perception, they say, "I can feel the burning in my brain, right here, moving down into my face, right here, like tiny ants crawling in there."

When a patient presents with several of these symptoms—it is a rare patient who has all of them—*Bartonella* (and mold toxicity) should come to mind. You have probably noticed by now that I keep including mold toxicity in this discussion, and for very good reason: although mold toxicity is caused by toxins and *Bartonella* is a bacterial infection, they create a remarkably similar release of inflammatory cytokines inside the body that sets off a chain of inflammatory reactions that cause all these symptoms. This means that their presentations are very similar, and they can be hard to distinguish. Worse, many of my patients have both *Bartonella* infection *and* mold toxicity. Because this chapter is about *Bartonella,* I will set aside references to mold toxicity, but I hope I have made my point.

The take-home message here is that *Bartonella* is uniquely capable of creating an inflammatory process that "wires" or overexcites the nervous system and immune system. This is one of the primary causes of profound sensitivity in an individual who is unfortunate enough to acquire this infection. The unusual constellations of symptoms described by my patients (often misinterpreted as psychosomatic by physicians who are unaware that there are medical conditions that can cause them) include intense reactions to stimuli such as light, touch, sound, smells, foods, chemicals, and electromagnetic fields. This same neurological sensitization creates symptoms that patients typically describe with such intensity that unless the physician understands the likely cause, they stretch the limits of credibility. To the uninitiated, a patient's description of a burning, crawling feeling in the brain sounds too weird to take seriously. To a physician who treats patients with *Bartonella,* however, it is a commonly encountered diagnostic clue that *Bartonella* is what this patient is dealing with. What someone else might view as a "crazy" description, I and other physicians who treat these kinds of patients see as a clear diagnosis. It helps to know what you are looking for.

Diagnosing Bartonella Infection

The best diagnosis of *Bartonella* infection is made clinically, by elucidating the symptoms that are typical of it. Unfortunately, the objective laboratory testing for *Bartonella* that is currently available is woefully inadequate. For a few "lucky" patients, the rather limited testing available does point to the diagnosis of *Bartonella*, but for most, the diagnosis is essentially empirical.

The most common test is a blood test that is an antibody titer for *Bartonella*, IgM or IgG. (A *titer* is simply a measurement of the reactivity of the immune system to a given infection. If it is positive to any extent, it suggests past exposure to that microbe, since the immune system has a memory of reacting to it. Unfortunately, it does not tell us whether the infection is current or old news.) We test for the two most prevalent species of *Bartonella, henselae* and *quintana,* in this manner. A positive test means only that the patient has been exposed to this species and has made antibodies to it. It does not confirm that the bacteria is present or active, but it gives us scientific evidence to consider *Bartonella* infection as a diagnosis. However, there are twenty-seven known species of *Bartonella* that can cause infection, and we have tests for only two of them. A negative test, then, does not rule it out, just as a positive test does not fully confirm the diagnosis.

Recently, a lab called Galaxy Diagnostics in North Carolina developed a culture combined with PCR technology that was hoped to be more accurate. (*PCR* refers to polymerase chain reactions, a technique used in molecular biology to amplify a few copies of DNA to make it easier to detect small amounts of substances in a biological specimen.) Interestingly, my colleagues on the East Coast find this test to be accurate 30 to 40 percent of the time—which is a breakthrough—but here on the West Coast, even in patients with a clear diagnosis, the test is rarely positive.

We have one other diagnostic approach that is more generally useful: the *Bartonella* challenge test. Several companies have created tinctures (which are typically alcohol or glycerin extracts of biological substances) that are specific for *Bartonella*. Using the A-Bart tincture from Byron White or the Bar-1 extract from Beyond Balance, we have a tool that has both diagnostic and therapeutic effects.

For this test, the patient is given five drops of one of these tinctures twice a day for two days. (For very sensitive patients, we start with less.) If there is no reaction whatsoever after two days, the dosage is increased to ten drops twice a day for two days. If still no reaction, the dosage is increased to fifteen drops twice a day. A patient who has *Bartonella* often will react immediately to a small dose of the tincture with an exacerbation of their typical symptoms. The exacerbation can be intense: a marked increase in fatigue, brain fog, pain, or anxiety. If a patient reacts in this fashion, we have the information we need, and the test is over: any reaction indicates the possibility of *Bartonella* infection. The reaction will usually subside in a day or two. It is important not to continue the test by increasing the dosage if the patient clearly has reacted to the tincture. Continuing the test can result in a severe and prolonged reaction, which must be avoided if at all possible.

This challenge test can be quite helpful in diagnosis given the lack of more objective testing. Again, however, a negative test does not completely rule out the possibility of *Bartonella.* A clinical diagnosis based on careful evaluation is what physicians must rely on here.

Treating Bartonella Infection

The comprehensive treatment of *Bartonella* infection is beyond the scope of this book. I will provide some general principles for treatment, but patients must find physicians who are knowledgeable and trained in this area.

Treatment is not for the dilettante. It requires an understanding of antibiotics and herbal supplements and especially *patience.* *Bartonella* is a deep-seated infection that is not easy to eradicate. It usually takes months, and sometimes years, to eliminate or even control. For physicians who are used to treating an infection with an antibiotic for seven to ten days and then being done…this is different. Several physicians have remarked to me, in reference to Lyme disease or *Bartonella,* that they know of no other infections that require this amount of consistency and vigilance, so what are we doing with this kind of prolonged treatment? I remind them that this type of treatment is standard for tuberculosis, so we have clear evidence that some infections do require a different approach.

LLMDs (Lyme Literate physicians)—the name given to physicians who are courageous enough to pursue treatment for their patients even when it is not fully sanctioned by the medical establishment—have tried to share information, but there is no absolute consensus on what constitutes the perfect treatment for *Bartonella* at this time. Treatment typically includes:

- **Antibiotics:** Rifampin, Septra, azithromycin (Zithromax), and Clindamycin are commonly prescribed. (Cipro is also used, but we try to avoid it because of the recently recognized incidence of chronic tendonitis as a side effect; see page 186.) Combining these drugs with hydrosol silver (Argentyn 23) has worked well. The use of antibiotics also must include the regular use of probiotics. The well-recognized effects of antibiotics on intestinal bacterial balance, referred to as the gut microbiome (see Chapter 12), require that we supplement with probiotics to prevent this disruption.

- **Herbal supplements:** The use of the A-Bart or Bar-1 tincture, described previously in the section on diagnosing *Bartonella,* is very helpful. In his excellent book *Healing Lyme Disease Coinfections,* Stephen Harrod Buhner outlines the herbal materials he has found most useful in treating *Bartonella.* His program, which is based on a deep understanding of how *Bartonella* affects the immune system, includes materials to reduce the inflammation created by inflammatory cytokines and support the organs and cells most affected by *Bartonella,* such as the spleen and red blood cells, along with herbs that have antibiotic properties.

  Lee Cowden, MD, has developed a series of herbal tinctures that are specific to *Bartonella.* His company, NutraMedix, provides them. I have found Banderol and Houttuynia to be particularly beneficial.

- **Homeopathic treatments:** These include a series of tinctures made by Deseret Biologicals that are specific to *Bartonella,* which I have found valuable.

- **Electrical treatments:** Variations on the Rife machine (see the sidebar below) and other similar treatments are available. Though sometimes helpful, these electrical treatments are often poorly tolerated by sensitive patients. I approach them with great care.

THE RIFE MACHINE

The Rife machine is a device that was developed by a researcher named Royal Rife between 1936 and 1950. Dr. Rife required a unique, high-powered microscope (no longer in existence) to analyze a patient's blood with an infection present. After hours of exposing the blood to tiny changes in electrical frequencies, he would observe a clear destruction of the invading microbe when an exact frequency of electromagnetic energy was applied. He then used very short treatments (no longer than six minutes each) of that exact frequency and found that doing so would resolve these infections within a few days.

The current use of Rife machines is based on supposedly knowing the correct frequency to use for a wide array of infections, including Lyme and *Bartonella.* Dr. Rife, however, did not find a single frequency for any organism; rather, only by direct observation under his microscope did he find the appropriate frequency for a specific patient. While some patients have been helped by the use of Rife machines, many others report Herx reactions, and the indiscriminate use of these frequencies has made many sensitive patients worse.

Many other treatments for *Bartonella* have been utilized with varying degrees of success. The ones I have referred to are simply the ones that have worked best for my patients. The use of lymphatic drainage, detoxification strategies, FSM, LENS, and osteopathic manipulation, all of which are discussed in more detail in Part 3, have been quite helpful in treatment. Other adjuncts to treatment will be covered in more detail in Part 3 as well.

Let me repeat that most patients require prolonged treatment to recover. It is not unusual for patients to improve with shorter treatments only to relapse. If mold toxicity also is present, it will be difficult to adequately treat *Bartonella* until the mold has been completely removed. I suspect that quite a few patients who are frustrated by regular relapses have not addressed the possibility of mold.

One last reminder: In a highly sensitive patient, the possibility of *Bartonella* as a major component of illness should be strongly considered. If it is not addressed, it will be difficult for the patient to get well.

How Bartonella Compares with Borrelia and Babesia

Now that I have emphasized the importance of evaluating and treating *Bartonella*, let me add some more complexity to the discussion.

While we frequently can make excellent and useful generalizations about *Bartonella* on the basis of very specific symptoms, sometimes those same symptoms appear to be a manifestation of *Babesia* and/or *Borrelia* (the microbe that causes Lyme) instead. Discussions with practitioners from around the United States suggest that what physicians on the West Coast perceive as clear manifestations of *Bartonella* sometimes look like *Babesia* to physicians on the East Coast. To make this even murkier, the immune systems of many patients often make rapid shifts in their focus from *Bartonella* to *Babesia* and back again. I have tried to encompass this concept by calling it "Babonella." My colleague Wayne Anderson, ND, used to call it the "*Bartonella/Babesia* dance." At other times, Lyme disease itself seems to manifest with the symptoms classically attributed to *Bartonella,* making it inherently difficult to diagnose on the basis of symptoms alone. But, alas, often that is the best we can do.

Therefore, let us spend a little time on understanding the typical presentations of the other components of Lyme. It is not unusual for me to see patients who have been treated by other LLMDs for long periods, with limited success, for symptoms that are more likely to be caused by a *different* infectious agent. For example, patients with intense anxiety and despair, gastric pain, and intense sensitivity to chemicals and light are much more likely to have *Bartonella* than Lyme, but their physicians may have been treating them for *Borrelia* for several years without seeing substantial progress. Knowing the most common symptoms manifested by each of these infections allows us to make better educated guesses as to what should be treated.

Symptoms Common to Most of These Infectious Agents

By "most of these infectious agents," I am referring to *Babesia, Bartonella, Borrelia* (Lyme), *Chlamydia pneumonia, Ehrlichia, Mycoplasma pneumonia,* Rickettsial infections, and viral infections. All of these infections are usually associated with fatigue and cognitive impairment and are often associated with swollen lymph nodes, sore throats, flulike symptoms, nondescript rashes, and sleep disturbances. So these symptoms do not help us distinguish one microbe from another as a causative issue.

Symptoms Often Present in Borrelia (Lyme) Infections

Lyme experts often distinguish two major forms of the disease on the basis of the major organ system that is presenting with symptoms: the nervous system *(neuroborreliosis)* and joint and muscle pain.

Fifteen to twenty-five percent of patients with *Borrelia* have neurological symptoms that reflect underlying inflammation of many aspects of the nervous system. These symptoms include:

- Meningitis (inflammation of the membranes surrounding the brain and spinal cord)

- Radiculitis (inflammation of nerve roots), which presents with localized nerve pain

- Peripheral neuropathy

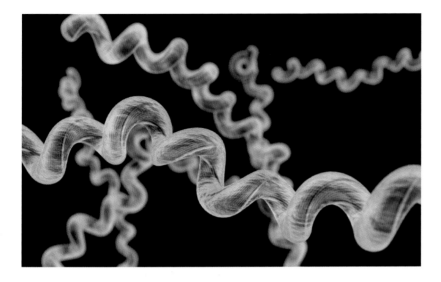

Figure 4.2. The spirochete of *Borrelia*.

- Cranial neuritis (inflammation of cranial nerves), which can include Bell's palsy (VII cranial nerve inflammation), double vision (III and/or VI cranial nerve inflammation), and/or optic neuritis (II cranial nerve inflammation)

- Headaches/neck pain

- Vertigo

- Encephalitis (inflammation of the brain)

- Myelitis (inflammation of the spinal cord)

The presence of any of these symptoms should alert a physician to the possibility of Lyme disease.

Joint and muscle pain is more diffuse and less specific, but 80 percent of patients with *Borrelia* develop muscle and/or joint pain, and 50 percent present with arthritis. While Lyme has been called "the new great imitator" (the first "great imitator" being syphilis, which, like *Borrelia,* is a spirochete; see Figure 4.2), neurological or joint symptoms are more common and should point the evaluating physician in that direction.

Cardiac symptoms, though rarer (occurring in 1 to 5 percent of Lyme patients), also are more typical of Lyme disease, so Lyme carditis (inflammation of the heart) may present with atrioventricular (A-V) blocks, which are arrhythmias (palpitations), or with pericarditis, chest pain, shortness of breath, or dizziness. In patients who present with cardiac symptoms, *Borrelia* (Lyme) is more likely to be the cause than any of the other co-infections we have been discussing.

Making the diagnosis of Lyme disease is fraught with difficulty.

The most common method of testing, called the Western blot test, essentially measures whether a patient's immune system has been exposed to *Borrelia* bacteria and is making antibodies to it. The results are expressed in numbers, called "bands," with each band representing a piece of protein present on the *Borrelia* bacteria to which the body is making an antibody. Here is where some knowledge about which bands are more relevant than others comes into play. There are quite a few nonspecific bands—namely 41, 58, and 66—that are common not only to *Borrelia* but to other bacteria and viruses as well. So, if those bands are present in a patient's test, they by no means prove the presence of Lyme. Band 31 also reacts with the Epstein-Barr virus (EBV), so it is not specific, either. Bands 18, 23–25, 34, 39, and 83–93 are more specific to Lyme disease; one or more positives in these bands are more diagnostic for Lyme but may not fully meet the CDC's requirement of five positive bands to make the "authorized" diagnosis. Those of us who work in this field have discovered that we need to take into account the detailed evaluation of symptoms, along with the patient's history and the results of the Western blot test, to make a tentative diagnosis of Lyme disease.

Now, let's delve into why this diagnosis can be so difficult to make on the basis of this test alone (which is considered by most to be the "gold standard"). First, it may depend on which laboratory runs the test. Unfortunately, in the late 1990s, a Lyme vaccine appeared that was hoped would help prevent Lyme disease. Unfortunately, it had to be pulled from the market quickly after it was discovered that the vaccine *caused* Lyme-like disease in about 20 percent of recipients. However, because this vaccine attempted to protect patients by using three specific protein pieces of the *Borrelia* bacteria, those protein pieces (bands) were left out of the standard testing performed by Quest Diagnostics and LabCorp, making their tests less accurate.

Many physicians, either untrained in the evaluation of Lyme disease or inclined to doubt its presence, use a Lyme antibody test, which is so inaccurate that it is of virtually no value whatsoever. Countless patients who thought they might have Lyme have been reassured by their physicians that they do not based on this inadequate test. The most comprehensive and accurate test, which LLMDs have relied on for years, encompasses the full range of bands and is run by the IGeneX lab.

While many other tests have appeared on the market, most of us who deal with these kinds of cases agree that none of them is as accurate or useful. Some of the newer tests are so much more sensitive that the results can be misleading and cause physicians to overdiagnose Lyme.

A different and more accurate assay has been run for the past five or so years by Advanced Laboratory Services. Rather than measure antibodies, it is a blood culture for *Borrelia.* The lab advertises a 92 percent accuracy rate for this test, and my experience corroborates that claim.

In addition to which lab runs the assay, there are several more significant problems. One is that there are so many species of *Borrelia,* and the Western blot test does not include them all. That test was standardized for *Borrelia burgdorferi,* our most common species, but recently it has been discovered that *Borrelia miyamotii* is far more prevalent than was appreciated. (More than 25 percent of ticks here in northern California carry it, as opposed to the 1 to 2 percent originally thought.) We also are discovering new species with regularity, and the available testing does not take these new species into account. Finally, in Europe and other parts of the world, still other species of *Borrelia* are more common. In Europe, for example, *Borrelia burgdorferi* is present, but so are substantial numbers of *garinii* and *afzelii* species, among others. Until testing includes *all* of these species, it will not be as accurate as we need it to be to help us make the correct diagnosis.

Another issue is that Lyme disease weakens the immune system so profoundly that it is often unable to make these antibodies. A negative test, then, in no way rules out Lyme disease. When there is a strong clinical suspicion that Lyme is present, treatment with antibodies will strengthen the patient's immune system, and he or she may well be able to make antibodies and have a positive test four to six months after beginning treatment.

One specific symptom that is considered pathognomonic (highly specific) for Lyme disease is the presence of a bull's-eye rash (see Figure 4.3) after a tick bite. Unfortunately, not all physicians, especially emergency room physicians, who may be the first to interact with these patients, are aware of this specificity. The presence of this rash alone is sufficient to make the diagnosis of Lyme disease and warrants six weeks of antibiotic therapy.

If a person is bitten by a tick, no matter how long it has been attached, the tick should be saved in a plastic baggie and sent to a laboratory to be evaluated for Lyme. If the test comes back positive, treatment should be offered; if it comes back negative, treatment is not necessary. Many state laboratories offer this service, and the specimen can be sent to IGeneX for evaluation as well.

> " The most comprehensive and accurate test, which LLMDs have relied on for years, encompasses the full range of bands and is run by the IGeneX lab. "

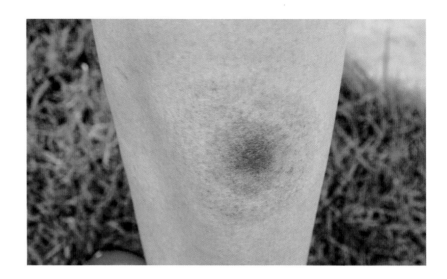

Figure 4.3. The characteristic bull's-eye rash caused by the bite of an infected tick.

> **"**
>
> **If a person is bitten by a tick, no matter how long it has been attached, the tick should be saved in a plastic baggie and sent to a laboratory to be evaluated for Lyme.**
>
> **"**

So there is your crash course in Lyme disease. To summarize the important takeaways for diagnosis:

- Making the diagnosis may be difficult, but *not* making the diagnosis can lead to years of disability, so it is worth the effort for anyone with an undiagnosed chronic illness whose symptoms match up with those commonly associated with Lyme to be tested as thoroughly as possible, despite the limitations of current testing options.

- Neurological symptoms—especially Bell's palsy or other unexplained neuropathies—may be a clear indication of the presence of Lyme disease.

- Unexplained joint or muscle pain should lead us to consider the possibility of Lyme disease.

- As discussed with *Bartonella,* if a neurologist or rheumatologist makes the diagnosis of "atypical" MS, ALS, Alzheimer's disease, Parkinson's disease, or rheumatoid arthritis, the possibility of Lyme (or mold or *Bartonella* infection) should be considered.

- Cardiac symptoms are more likely to represent Lyme disease than any of its co-infections.

Symptoms Often Present in Babesia Infections

Babesia is a malaria-like parasite that infects red blood cells. It is usually acquired from a tick bite. The symptoms may begin with a flulike clinical picture (as is the case with the other tick-borne illnesses), or symptoms may not emerge until long after the tick bites, when the person's immune system has been weakened or compromised.

The most prominent symptoms of *Babesia* infection include:

- Sweating, especially at night
- Shortness of breath or "air hunger" (a feeling of being unable to take a deep breath even though oxygen saturation measurements are normal)
- Frontal head pressure or headaches
- Extreme cognitive dysfunction
- Intense psychological symptoms, including hallucinations or a sense of derealization
- A disconnect between the intention to perform an action and the ability to do so, such that accomplishing simple everyday actions may require a great deal of focus

While other symptoms may be present, these symptoms are pathognomonic for *Babesia* and strongly indicate its presence. However, these symptoms can also be found in patients with mold toxicity and at times in the presence of *Bartonella*.

Confirming the diagnosis of *Babesia* can be quite difficult. It is sometimes possible to detect *Babesia* by looking at a thick smear of blood under a microscope, but this approach is unreliable after the infection has been present for more than two weeks—which is almost always the case. Blood testing that uses PCR technology (see page 88) can be helpful, but currently only two species of *Babesia* (*duncani* and *microti*) can be tested, and more than 100 other species are known to infect humans. The FISH (fluorescent in-situ hybridization) test has the same limitations, as do the blood tests for antibodies to *Babesia* that are available. When positive, these tests can help us determine whether to include *Babesia* in the diagnostic assessment of a patient, but when negative (which is usually the case), they do not rule it out. As you can see, testing is not very accurate, and, as with Lyme and *Bartonella,* we are often placed in the position of having to treat *Babesia* empirically, without the firm laboratory backing we would prefer.

Quite a few other co-infections may be present, but these are the major ones we work with routinely.

I hope I have clarified the difficulties in making a precise diagnosis and emphasized the importance of looking at each patient's symptoms in detail. At the risk of oversimplifying things, I have summarized the key symptoms in the table below. Keep in mind that many symptoms overlap and that microbes and humans do not always behave in predictable ways when they interact. Also keep in mind that mold toxicity, which is often present along with Lyme disease and its co-infections, has very similar symptoms, which means that teasing these issues apart is quite an art (and we have only a little bit of science to help us). This is merely a guide, but I hope it's a useful one.

Key Symptoms That May Help Distinguish Bartonella from Borrelia and Babesia

Bartonella	anxiety, sometimes with panic attacks; depression, sometimes with feelings of hopelessness and despair; derealization; pain on the soles of the feet, especially upon waking; visual disturbances; headaches at the back of the head; increased sensitivity to light, touch, smells, foods, chemicals, or EMF; a sensation of internal vibration; pseudo-seizures; dystonias; muscle twitching and paresthesias; pelvic or bladder pain
Borrelia	bull's-eye rash, cranial neuropathies (Bell's palsy, double vision), peripheral neuropathy, paresthesias, cardiac pain or arrhythmias, intense joint pain that may migrate, global headache (whole head hurts)
Babesia	sweating, especially at night; headache or pressure at the front of the head; shortness of breath or "air hunger"; neurological "disconnect"; severe cognitive impairment; severe psychological symptoms

With this guide, we can assess a patient carefully, make an educated guess as to which infection (or toxin) is the biggest threat, and begin treatment there. Given how inherently complicated this effort is, treatment should be instituted and followed only by those physicians who are well trained in how to manage these interwoven infections and toxic responses.

CHAPTER 5

Mast Cell Activation Syndrome

Yes, There Is Another Elephant in the Room

Just when you thought things couldn't get much more complicated, along comes mast cell activation. Dealing with profound, interrelated, and persistent inflammation that affects literally all of the systems of the body seems like more than enough. Indeed, the complexity of mold toxicity coupled with a wide variety of chronic infections is more than most physicians are willing to take on.

Patients with mast cell activation often present with a wide array of symptoms that fluctuate in unpredictable patterns, such as:

- Intense anxiety and depression
- Severe pain that can localize to joints, muscles, tendons, and/or bones
- Unusual neurological symptoms, like numbness and tingling in different parts of the body, paralysis, and pseudoseizures
- Headaches
- Ringing in the ears (tinnitus)
- Sensitivity to a wide variety of stimuli, such as light, touch, sound, smells, foods, chemicals, and electromagnetic fields (EMF)
- Sore throat
- Swollen lymph glands
- Indigestion, including diarrhea, constipation, bloating, gas, distension, and heartburn

- Chronic debilitating fatigue

- Insomnia

- Cognitive difficulties, including brain fog and decreased focus, memory, and concentration

- Pelvic pain

- Interstitial cystitis (a painful inflammation of the bladder)

- Shortness of breath

- "Air hunger" (see page 97)

- Skin rashes

- Difficulties with equilibrium and balance

The most obvious response to someone who presents with such a broad array of symptoms is to think, "No one could have *all* of these symptoms; they must be in your head." It should not come as a surprise that the vast majority of these unfortunate individuals are immediately treated as if this litany is psychosomatic and dismissed with a prescription for an antidepressant or anti-anxiety drug with a clear message: "I can't help you. You need a good psychiatrist." And most of these patients, though they personally doubt this diagnosis, cooperate with their physicians and do see psychiatrists, who duly provide medication with minimal benefit. (I am sorry to belabor this description, but it is so common that it is, unfortunately, the rule rather than the exception.)

Years pass. The patients do not get much better; in fact, they often get worse. If they are lucky, they find Lyme Literate physicians (LLMDs) who tell them that yes, there is something that can explain these symptoms, and treatment for Lyme disease often helps to varying degrees. With a little more luck, the physician is familiar with mold toxicity, which also can explain these symptoms, and with treatment for mold, the patient gets better still and sometimes re-covers completely. But some patients remain so sensitive that every treatment they try—be it a pharmaceutical, an herbal supplement, a homeopathic remedy, acupuncture, or electrical stimulation—makes them feel worse.

The good news is that we have recently realized that there is another elephant in the room: mast cell activation. (In Chapter 6, I will discuss two more elephants, porphyria and carbon monoxide poisoning, but let's not get ahead of ourselves.) When doctors under-stand what mast cell activation is and, most importantly, diagnose it, additional treatment can be provided that allows suffering patients to improve and to move forward with other treatment.

What Is Mast Cell Activation?

Let's start with these questions: what are mast cells, and why should we care about them becoming activated?

Mast cells (see Figure 5.1) are an integral component of our immune system. Though not present in large numbers, as are the better-known white blood cells (such as neutrophils, lymphocytes, and monocytes), mast cells function as a critical *bridge* between the immune system and the nervous system. This is not merely a metaphoric connection, but an actual, structural bridge that physically connects the two systems.

I find it fascinating that many mast cells in the body are connected, both directly (by abutting on nerve cells) and indirectly (by releasing chemicals), to the nervous system. This implies that mast cells are vital coordinators between the two systems. Their basic function (among many others) is to coordinate the immune system's activities that deal with toxins and infectious agents. Ah, so the very agents that we have been focusing on in this book require functional mast cells to engage the body's natural defense systems!

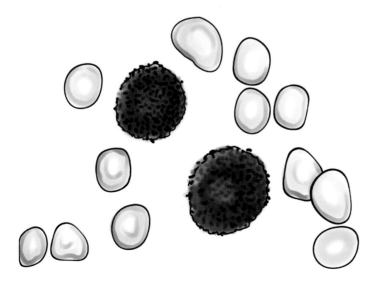

Figure 5.1. This is what mast cells look like under a microscope when stained by a conventional laboratory. Note the purple dots, which are granules, most of which contain histamine that can be released at a moment's notice.

Mast cells are present in every tissue of the body, but they are found in the largest numbers in those tissues that come into closest contact with the outside world. That makes sense; if you are going to coordinate dealing with toxins and infections, then that coordination should start with those tissues that will be in the most immediate and direct contact with those toxins and infections—namely the tissues that line the sinus, throat, gastrointestinal tract, respiratory tract, skin, and genitourinary tract.

Mast cells are capable of making more than 200 biochemical mediators (substances that convey signals to other cells and tissues to alert them to incoming problems) in response to a wide variety of stimuli. When you view mast cells under a microscope, what draws your immediate attention is that they are loaded with tiny round dots, called *granules.* Some of these granules contain preformed materials to initiate these signals quickly—the main ones being histamine, serotonin, and tryptase. While all of these mediators can be important, histamine is the most obvious in terms of what it can do to the body if inappropriately stimulated.

So, when a mast cell is *appropriately* stimulated by a toxin or an infectious agent, it can immediately release some of these granules, which move quickly through the bloodstream to coordinate a proper immune response. However, when a person's immune system is overloaded with toxins or infection and his or her personal biochemistry and genetics align to create a "perfect storm," the mast cells become *activated.* Other words that can help us understand activation include hyperreactive, overexcited, fragile, and trigger-happy. See Figure 5.2 for a microphotograph of what an activated mast cell looks like.

What I am hoping to convey here is that when mast cells become overly reactive, they lose the specificity of their response and can start reacting to ordinary stimuli that normally would not engage their attention. Now they react to foods that were easily consumed before, or to light, touch, sound, smells, or chemical exposures, triggering the sudden release of histamine and other substances that can wreak havoc in the body within mere moments. To put it simply, the body becomes sensitized to such an extent that *almost anything* can cause an unpleasant reaction. I have seen patients react to drinking a glass of water with an intense histamine response.

Resting mast cell **Activated mast cell**

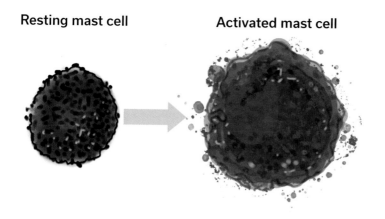

Figure 5.2. Here, you can see the difference between a normal mast cell and one that has become activated.

Uninformed family members, friends, acquaintances, and even physicians look at these suffering patients and think: "That's impossible. No one can react to drinking a glass of water." But they can. And they do. It seems almost crazy, I know, but there is a true cause of this reaction, and it is not psychological. What we have is an individual who has become so reactive that he or she does not know which direction the next assault is coming from. These patients often resist eating or drinking anything because they cannot be sure of what will have a debilitating effect on their bodies.

I want to emphasize that mast cell activation is a real, *physiological* process. And, as I hope you can immediately see, it is frightening, chaotic, random, and very hard to deal with.

Here comes the first kicker: mast cell activation is often triggered by mold toxicity and/or infections like *Bartonella* and Lyme disease, as well as a wide variety of viral infections. Many patients have a genetic predisposition to mast cell activation, but whether or not it ever manifests in their lifetime depends to a large extent on their exposures and how well their immune systems function.

And here comes the second kicker: it is not rare. While we have only recently begun to understand it, it is estimated that mast cell activation may be present to some extent in up to 10 percent of the population. I would estimate that 50 percent of my ultrasensitive patients have a mast cell activation component. It is therefore important to understand how mast cell activation may reveal itself and what can be done to treat it.

Finally, keep in mind that mast cell activation is not the same as food allergies. Food allergies are consistent: the same food always provokes the same response. Reactions to foods caused by mast cell activation depend greatly on the state of activation. During a flare-up, virtually any food can provoke a reaction, and during a dormant period, the same food may be well tolerated. As you can imagine, this is both confusing and frustrating for patients (as well as their allergists and other doctors) because they often can't get a good handle on what they can and can't eat. This very point, in fact, helps us view these variable reactions as suggestive of mast cell activation.

Mast Cell

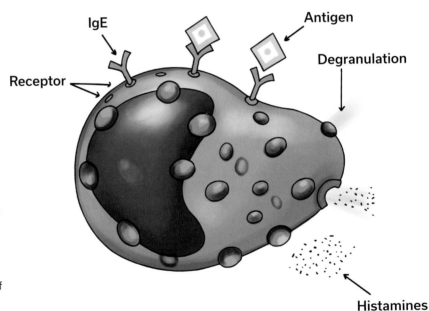

Figure 5.3. Histamine receptors sit on the membrane of each mast cell, ready to react to a variety of stimuli. An antigen can be an allergic substance or literally anything that the activated mast cell finds "annoying." When the receptors respond, the cell releases preformed granules, many of which contain histamine.

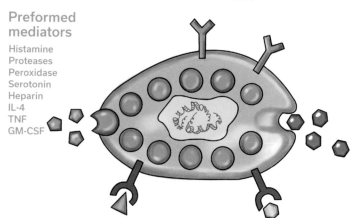

Physical Activators

Heat   Cold   Pressure

Preformed mediators

Histamine
Proteases
Peroxidase
Serotonin
Heparin
IL-4
TNF
GM-CSF

Newly synthesized mediators

IL-1,2, 3, 3, 5, 6,
7, 8, 10, 12, 13,
15, 18, 21, 23
Prostaglandins
Leukotrienes
VEGF, NGF, FGF
PAF
TGF-beta
CCL3
CCL2
IFN-alpha, -beta, -gamma
CRH
TSLP
CXCL10
TNF

Receptor-binding agonists

IgE + Ag   Complement hormones
Monomeric IgE   Neuropeptides
IgG + Ag   Cytokines
Ig light chain   Microbial products

T and B cell ligands

B7   CD40L
PD-L1   CCL19
OX40L   4-1BB
CD30L

Figure 5.4. These are just a few of the 200+ mediators that are made by mast cells and are released under a variety of circumstances. These mediators are grouped into commonly encountered families of biochemical substances.

Symptoms of Mast Cell Activation

If you are wondering why we are just now discovering how prevalent and important mast cell activation is, let me take a few moments to explain. Given individual genetics and biochemistry and the fact that mast cells can produce more than 200 different biochemical mediators depending on which stimuli they are exposed to, mast cell activation can produce an astonishing array of symptoms—so many, in fact, that, like the diagnoses of mold toxicity and Lyme disease, physicians have been missing this diagnosis for years because they could not wrap their heads around the varied ways in which it can present itself. I also believe that we are seeing an epidemic of it because we have not yet recognized how toxic our world has become; this, too, has sensitized many individuals to a greater extent than we are prepared to acknowledge.

In his groundbreaking book *Never Bet Against Occam: Mast Cell Activation Disease and the Modern Epidemics of Chronic Illness and Medical Complexity,* Lawrence B. Afrin, MD, lays this out in wonderful detail. I would like to utilize his method of taking the body system by system and showing how mast cell activation can affect each one, but I will try to do so in a somewhat simpler format than his. I also want to emphasize that symptoms can be chronic or can flare up for no obvious reason. Because histamine release is a common result of mast cell activation, I will mention some of the more obvious symptoms first, as these begin the process of drawing doctors' attention to the diagnosis. Dr. Afrin's books and articles offer much more comprehensive listings, but the following symptoms are the ones that I have seen most often in my sensitive patients:

- **Gastrointestinal:** Reactions that occur *quickly* after eating or drinking *anything,* especially if they involve something that appears allergic in nature, such as flushing, sweating, tachycardia (rapid heartbeat), itching, swelling of the tongue, or wheezing. Abdominal bloating, gas, diarrhea, pain, and nausea (often with vomiting) are commonly reported. These issues that clearly develop within minutes of oral exposure are the most obvious tip-offs that mast cell activation may be the culprit.

- **General:** Malaise, fatigue, temperature dysregulation, and weight loss or gain

- **Sensitivities:** Increased sensitivity to light, touch, sound, smells, foods, chemicals (multiple chemical sensitivities), and even electromagnetic fields (EMF)

- **Lungs:** Wheezing, shortness of breath, "air hunger," laryngitis, and bronchitis

- **Skin:** Rashes of every description, especially itchy ones (pruritus), often triggered by taking a hot shower and often associated with flushing and hives

- **Sinus/Oral:** Postnasal drip, a tickling sensation in the throat, congestion, sinusitis, rhinitis, and pharyngitis

- **Cardiovascular:** Light-headedness, weakness, dizziness, vertigo, even syncope (fainting), palpitations, arrhythmias, chest pain, areas of swelling (edemas) that move around the body, and high blood pressure. POTS (postural orthostatic tachycardia syndrome) is also common.

- **Pelvic:** Pelvic pain, bladder pain, flank pain, cystitis, vaginitis, prostatitis, and unexplained inflammation of the pelvic region

- **Musculoskeletal:** Diffuse, shifting muscle and joint pain (resembling fibromyalgia) that often responds poorly to the usual pharmaceutical treatments

- **Neurological:** Headaches, paresthesias, tremors, tics, seizures, pseudoseizures, and dysautonomias (unusual and uncontrollable spasmodic, twisting, writhing motions)

- **Psychiatric:** Anxiety, often with panic attacks; depression; unpredictable mood swings; cognitive difficulties, including "brain fog" and difficulties with focus, memory, and concentration; and insomnia

- **Visual:** Eye irritation (often described by patients as burning, sandiness, or dryness), loss of visual clarity, and trembling of the eyelids (called *blepharospasm*)

This is an impressive array of symptoms, yes? And keep in mind that I have not included the entire gamut of symptoms known to be associated with mast cell activation in an attempt to keep it as "simple" as I can.

I would like to emphasize that although many of these symptoms appear to be allergic in nature, there is a *distinct difference* between immediate allergic events, which are mediated by IgE antibodies, and the sudden release of histamine and other mediators from activated mast cells. I think it is important to emphasize this difference to help us consider the possibility of mast cell activation when the knee-jerk response is to assume that what is occurring is an allergic reaction.

To give an example, I was once attempting to administer intravenous phosphatidylcholine to a very sensitive patient. Within seconds, she felt "awful," her heart rate jumped to 112 (the normal resting range is 60 to 100 beats per minute), and she broke out in a sweat and felt itchy all over. Some physicians would assume that this patient was having an allergic reaction to the phosphatidylcholine, but actually it was caused by the sudden release of histamine from mast cell activation. While the release of histamine is a part of the activation of IgE-mediated allergic reactions, it is helpful to attempt to differentiate the two responses. For this patient, starting with an administration of Benadryl rather than an injection of epinephrine (which would be a later strategy) was very helpful.

Diagnosing Mast Cell Activation Syndrome

I have good news and bad news. The good news is that there are effective ways of treating mast cell activation, which I will get to shortly. The bad news is that making the diagnosis can be quite tricky—and at times almost impossible.

There are several reasons for this difficulty in diagnosis. First, I have emphasized that mast cells can release more than 200 different biochemical mediators under different conditions, meaning that a vast array of things theoretically could be measured to confirm this diagnosis. However, when mast cells are activated, the release of these biochemical mediators is often so rapid that they are in and out of the bloodstream within minutes. Their effects may last much longer, but catching elevated levels with a blood test is difficult indeed. Multiple measurements may need to be taken, especially when patients feel their worst, to catch these evanescent spikes. Furthermore, many of these mediators cannot be accurately measured by routine testing and require specific cold centrifuges (not available at most laboratories) and immediate processing for measurements to even begin to be accurate.

Having said this, the following tests have been shown to be of value in clarifying the diagnosis of mast cell activation syndrome:

- **Total serum tryptase:** A positive test would be above baseline or "normal" during or within four hours of a symptomatic episode.
- **Chromogranin A**
- **Plasma heparin and/or histamine:** Dr. Afrin has shown that using a cold centrifuge to measure histamine makes this testing much more accurate.
- **Urinary N-methylhistamine**
- **Urinary PGD2 or its metabolite, 11-beta-PGF2-alpha**
- **Leukotriene E4**
- **Antibodies to IgE (anti-IgE IgG) and antibodies to IgE receptors** are not currently seen as being diagnostic of mast cell activation syndrome, but rather a hint that autoantibody-mediated mast cell activation may be present.
- If a biopsy of tissue has been obtained, it can be particularly beneficial to examine or re-examine that biopsy after staining it with CD 117 to look specifically at the mast cell population in that specimen.

- Treatments are especially relevant for patients who are genetically negative for KIT-D816V. That particular gene is strongly associated with the rare condition known as *mastocytosis,* and if a patient is negative for that gene, he or she may be a candidate to try a variety of medications, including imatinib. The use of these medications is limited to those who specialize in hematology; they are not generally available to most physicians. I do want to make clear that this information about mast cell activation must be separated from the extremely rare blood cancer called *mastocytosis,* which is not a part of this discussion.

Even if all of these tests have been done repeatedly, a physician may not be able to make a definitive diagnosis of mast cell activation. However, if the treatment methods outlined on the following pages are successful in alleviating a patient's symptoms, then there is strong presumptive evidence that this is indeed the right diagnosis.

Treating Mast Cell Activation Syndrome

Many strategies are available for the treatment of mast cell activation, and combining strategies often improves outcomes. I would like to stress again that each patient is unique and that there are no clear rules or algorithms that can be universally applied. It is important for patients to remain hopeful and enthusiastic and to continue trying new treatments even when previous ones have set them back. Eventually, you are bound to find treatments that work.

Basic Principles of Treatment

Let's jump right into how to approach treatment. There are several basic principles around which I will outline specific treatments. Understanding those principles allows us to make sense of what comes next:

- Reduce the production of biochemical mediators from the mast cells. Most important here is to identify what is triggering the activation in the first place. In my experience, mold toxicity is the most common trigger, followed by *Bartonella* infection, which often is a co-infection of Lyme disease. I have found that when the trigger is treated correctly, mast cell activation often disappears.

A variety of desensitization therapies are helpful in this regard. While identifying and treating the trigger(s), it is extremely helpful to quiet mast cell activation by using what are called *mast cell stabilizers.*

- Interfere with the mediators that are released so that they have a less potent effect on the body.
- Counter the unavoidable effects of the released mediators.
- Consider the use of a low-histamine diet.

Diet and Exercise

You can find a bewildering array of low-histamine diets by surfing the web. The essence of these diets is limiting the oral intake of histamine, which clearly would add to the reactivity of a mast cell–activated patient. Histamine is often produced when food, especially protein, sits uneaten. This means that leftovers, overly ripe or fermented foods, ready-cooked meals, and canned foods, along with smoked meats, pickled foods, vinegar, and nuts, are more likely to be high in histamines and should be avoided.

I have found that about 50 percent of my patients with mast cell activation respond well to a low-histamine diet, and the other half do not. Therefore, I encourage patients to go online, find a low-histamine diet that they believe they can work with (these diets are quite restrictive and can be difficult to follow), and try it for two full weeks. If they experience no obvious improvement, there is no reason to continue the diet.

In a high percentage of my patients, mast cell activation is triggered by mold toxicity; therefore, most of them need to be on a low-carbohydrate diet. Carbohydrates, especially simple sugars, provide nutritional support for mold and *Candida,* so carbohydrate intake must be reduced as much as possible. The majority of my patients have already discovered, before they come to see me, that avoiding sugars and other carbohydrates is helpful.

The main point I want to make about exercise is that in some people it can provoke a worsening of mast cell activation symptoms. This is especially true in those who experience post-exertional malaise, a condition in which there is a clear worsening of fatigue and muscle pain after exercise that can last for hours or even days. Post-exertional malaise will not improve with persistent attempts to exercise, even when the individual is encouraged to do so by well-intentioned family members and friends. It is a measure of energy reserves, which most of my patients simply don't have. So, although I do encourage exercise, it is important to monitor how much exercise can be done without provoking a worsening of symptoms.

Supplements and Medications for Mast Cell Activation

Using the basic principles of treatment listed at the beginning of this section, we can select different families of supplements and medications to help stabilize mast cells and prevent the histamine that is being released from having as profound an effect. Usually, these materials are synergistic, meaning that using several of them, especially ones with different modes of action, can improve a patient's response significantly. In my clinical experience, patients tend to respond better to either medications or supplements, but not both.

I often start with three easily obtained materials to gauge a patient's response or reactivity. How he or she responds (or reacts) informs me as to what to try next. If the patient responds to a medication, I focus on different pharmaceuticals to facilitate treatment. If he or she responds to natural supplements, I focus on those instead. If the patient responds to both types of materials, I expand our treatment efforts into both areas.

Unfortunately, we have very little science or testing to help us know which materials will work for which patients, so it essentially comes down to trial and error. The better a patient understands this, the easier it is to cooperate with treatment attempts and to be prepared for the possibility that any material could exacerbate symptoms.

Quercetin

I usually begin with quercetin, a natural material that is in the vitamin C family. It is an excellent mast cell stabilizer. Most of my patients tolerate it well, but a few—around 15 percent—react badly to the usual dosage, which starts at 500 milligrams once a day. (It is important to take quercetin thirty minutes before a meal because then it can begin the process of stabilizing the mast cells before they can be adversely affected by eating.) If 500 milligrams is well tolerated for several days to a week, I encourage the patient to slowly work up to four 500-milligram doses taken thirty minutes before each meal and at bedtime. If this dosage is well tolerated and helpful, it can be doubled. For those patients who react negatively to the initial dosage of quercetin, I try the product NeuroProtek LP, which contains only 40 milligrams of quercetin and sometimes is better tolerated.

Loratadine (Claritin)

Quercetin helps stabilize mast cells so that they become less reactive, but it does not do this job perfectly. Excess amounts of histamine are likely to be released, and it is helpful to use both H1 and H2 histamine receptor blockers so that the released histamine will have less of an adverse effect. I usually start with 10 milligrams of Claritin, which is easily obtained over the counter. I recommend taking the first dose at bedtime because a few patients find that Claritin makes them sleepy. If this goes well, the patient can try taking it again in the morning. Again, a few patients find that Claritin makes their symptoms worse. Dr. Afrin has discovered that the worsening often is not from the loratadine itself, but from the fillers mixed into the capsule or pill. Some patients who react badly to Claritin do much better when the loratadine is made up in a purer form by a compounding pharmacy.

Famotidine (Pepcid)

Once we have determined that Claritin is helpful (improvement may be immediate or could take up to two months), I move on to Pepcid, an H2 histamine receptor blocker, starting with 20 milligrams taken once daily at bedtime and increasing the dosage to twice daily if well tolerated.

About 50 percent of my patients report improvement after taking one or more of these three materials, ranging from slight to major. Again, their reactions help inform me about what to try next. As noted, some patients do much better with natural materials, some do better with pharmaceuticals, and some respond to a mix of both. If a patient reacts negatively to pharmaceuticals, I focus my next efforts on natural materials. Conversely, if a patient reacts poorly to quercetin, I might focus more on pharmaceuticals. Each patient is unique. Once we have clarified the category of treatment to which an individual responds best, we can focus on that category in more detail.

Natural Substances That May Be Helpful

This is not intended to be a complete list, nor am I intending to focus on any particular manufacturer, but I will provide specific product names of the supplements that I have found to be of particular benefit for my patients. The materials noted here represent my personal observations; this list is not intended to be comprehensive, but merely a starting point for further treatment efforts. The dosage for each should be adjusted in response to how well it is tolerated.

- **Perimine (extract of perilla seed)** is a blend of bioflavonoids, including luteolin and rosmarinic acid, that has been helpful for some patients. Starting with one capsule taken thirty minutes before a meal, the dosage can be increased to one capsule before each meal.

- **HistDAO** contains the enzyme diamine oxidase, which helps break down histamine, supporting its healthy degradation. Starting with one capsule taken thirty minutes before a meal, the dosage can be increased to two to three capsules before each meal.

- **AllQlear,** derived from quail eggs, contains a tryptase blocker (tryptase being another mediator released by mast cells). Starting with one chewable tablet taken thirty minutes before a meal, the dosage can be increased to two capsules before each meal.

Pharmaceuticals That May Be Helpful

- **Ketotifen** is an H1 histamine receptor blocker, a mast cell stabilizer, and a functional leukotriene antagonist that comes at the process for treating mast cell activation from multiple perspectives, making it helpful in a variety of ways. It is available by prescription from compounding pharmacies. I usually start with a small dosage, often one 0.5-milligram capsule taken at bedtime, and slowly increase it as tolerated. It is also available as ketotifen fumarate eye drops for patients with eye-related symptoms, with the typical dosage being one drop in each eye two or three times a day.

- **Cromolyn sodium,** another excellent mast cell stabilizer, comes in a variety of forms. It has long been used in inhaler form at a dose of 20 milligrams to help with exercise-induced asthma; the oral form is available in 100-milligram vials. Sometimes it is helpful to start with a tiny amount of the oral form and gradually increase the dosage as tolerated. Cromolyn sodium is also available as an ophthalmic solution for patients with eye symptoms.

As noted earlier, there are additional treatments and medications that can be prescribed only by hematologists and allergists and are beyond the scope of this book. For details, I refer you to Dr. Afrin's excellent discussion of their use in his books and published papers.

Summary

We are just beginning to appreciate how widespread mast cell activation syndrome is. It should be considered in the diagnosis of every patient who presents with an array of symptoms from many organ systems, and especially in patients with mold toxicity and Lyme disease, since I find these to be common triggers for mast cell activation.

So, first of all, think about mast cell activation as a diagnosis. Second, remember that it is treatable, so there is no need for patients to feel helpless or hopeless as they suffer with a complex, debilitating illness. Seek out physicians who are aware of and knowledgeable about mast cell activation to achieve optimal benefits.

Now that I am more aware of this clinical entity, I find mast cell activation syndrome in perhaps half of my patients who present with unusual sensitivities or reactivities. Getting these patients started on treatment can be very difficult, as they are already overly reactive to almost everything. Addressing mast cell activation early in the treatment process, before starting more specific treatment for mold toxicity or *Bartonella*, often helps quell that reactivity and allows the patients to begin treatment with a much greater chance of being able to tolerate it. In sensitive patients, mast cell activation (if present) significantly interferes with the ability to tolerate the treatments that will cure what is causing it by dramatically increasing their sensitivities. For this reason, looking for and treating it early on is often essential to quieting down that reactivity and allowing the patient to move forward. While logically you might think, "Let's just fix the mold, and then we won't have to even think about mast cell activation," it is also true that unless the mast cells are quieted down, it may be impossible to even begin to address the mold because the patient will not be able to take even tiny doses of the necessary binders.

CHAPTER 6

Porphyria and Carbon Monoxide

Still More Elephants in the Room?

I have been interested in and studied porphyria on and off for perhaps thirty years, and I have a confession to make. Many physicians have suggested that porphyria is more common than is currently appreciated. Accordingly, I delved into it and periodically test some of my patients for its presence. A larger percentage than I might have anticipated have tested positive for porphyria, but then I got mired in the difficulties of treating it and integrating those treatments into the larger therapeutic plan. Somewhat overwhelmed by these difficulties, I abandoned the subject, only to be re-intrigued every few years and to start again.

My most recent interest was piqued by the new book by Steven Rochlitz, PhD, called *Porphyria: The Ultimate Cause of Common, Chronic & Environmental Illnesses.* I am grateful to Dr. Rochlitz for his extensive work on this subject, which helped me finally grasp enough of it that I feel I can broach it here. What appears in this chapter is by no means to be taken as gospel. I believe it is important for this information to be included in our understanding of the conditions that can contribute to hypersensitivity, but—spoiler alert!—I still have not figured out how to integrate the treatment of it into the bigger picture. Despite this, I think porphyria deserves its own chapter, and I hope that others will find ways to do a better job of integration than I have.

Porphyria is a complicated subject. My goals here are to help you begin to understand it (admittedly, with simplifications) and to help place our understanding of it into perspective. As already noted, many medical writers have recognized that porphyria may be more common than is usually appreciated, but efforts to diagnose and treat it have waxed and waned, probably due its unique complexity and to the fact that the treatment for porphyria is in some ways in direct opposition to the treatment for some of the causes that trigger it.

What Is Porphyria?

So, first off, what is porphyria? Let me begin by reminding you that our bodies can utilize the red blood cells we make for only about ninety days, and then they have to be recycled to make new red blood cells. One of the central components of red blood cells is heme (think hemoglobin). For a wide variety of reasons, if the liver is not working properly and is genetically predisposed to a deficiency in making certain enzymes, the liver may be unable to make an adequate quantity of the enzymes required for heme to be properly recycled. The result is a buildup of some of these breakdown products, called *porphyrins.* When porphyrins accumulate in the body, we get porphyria.

There are two main types of porphyria: *primary* and *secondary.* The primary porphyrias are rare genetic abnormalities with eight different names, depending on which specific porphyrin is in the greatest excess. I have never seen a case of primary porphyria in my clinical experience, and that is not what I am going to discuss here. However, *secondary* porphyria, meaning a porphyria that is created by an infectious process or toxin, is not a rare occurrence. In fact, I think it is much more common than is generally appreciated.

Symptoms of Porphyria

The symptoms of porphyria are global, and they are similar to those of mold toxicity, Lyme disease and its co-infections, mast cell activation, and PANDAS. Like these conditions, porphyria sets off a series of biochemical reactions that are inflammatory in nature and can affect virtually every system of the body. The symptoms that I find most useful for drawing attention to porphyria as a contributing factor to chronic inflammatory-type illnesses are especially psychological in nature. They include intense anxiety, depression, and panic attacks, along with severe nausea and vomiting. When a patient experiences these symptoms shortly after taking a new supplement or medication, and the reaction lasts longer than two or three days (a typical Herx reaction), it suggests to me that a porphyric reaction has been triggered by or is secondary to the ingestion of that material. In my clinical practice, I commonly see the onset of severe anxiety and panic, often with intense nausea and vomiting, which goes on for many days or weeks after beginning a patient on a new antibiotic or supplement.

Other symptoms of porphyria include palpitations, weakness, unusual neurological symptoms, paresthesias, peripheral neuropathy, abdominal pain, constipation, diarrhea, allergic reactions, asthma, shortness of breath, chest pain, insomnia, skin rashes, dizziness, and increasing intolerance to medications or supplements, stress, light, sound, heat, cold, odors, and electromagnetic fields (EMF).

Primary Symptoms of Porphyria

PSYCHOLOGICAL	GASTROINTESTINAL	NEUROLOGICAL
• Anxiety • Panic • Obessive compulsive disorder (OCD) • Irritability • Restlessness • Paranoia	• Nausea and vomiting • Abdominal pain or symptoms after eating (like MCAS)	• Paresthesias • Tremors • Burning sensations • Abnormal sensations • Tics

Other Symptoms of Porphyria

- Tachycardia, hypertension (labile)
- Fatigue and weakness
- Dysuria and pelvic pain
- Shortness of breath, bronchospasm
- Edema
- Dizziness, visual symptoms, confusion
- Sun sensitivity
- Rashes, especially after sun exposure; pruritis
- Insomnia
- Fibromyalgia-like presentation
- Chest pain
- Increased sweating
- Peripheral neuropathy
- Hallucinations or seizures

Diagnosing Porphyria

Similar to the difficulties of diagnosing mast cell activation is the underlying theme that porphyrins are present in the body for very short periods (think of them as evanescent or transitory), although their effects can last for weeks. This means that a single test, like a snapshot, may not be sufficient to prove their presence in a patient's body. To optimize testing, urine must be collected when the patient feels the absolute worst (when the porphyrins are most likely to be present). As soon as the patient collects the urine, the specimen container must be wrapped in aluminum foil (because porphyrins are very sensitive to light) and the specimen placed in the refrigerator until it can be brought to a laboratory for analysis.

Most labs, including LabCorp and Quest, can run this test easily. Any elevation of any porphyrin in the specimen suggests the possibility that the patient is experiencing a secondary porphyria.

Treating Porphyria

In my experience, like mast cell activation, secondary porphyrias are most commonly triggered by mold toxicity and Lyme disease (especially *Bartonella*) but can also be triggered by viral infections, *Chlamydia* and *Mycoplasma* infections, and other toxins. This means that it is important not to lose sight of the need to treat the primary trigger or cause of the porphyria while addressing the biochemical effects of the porphyria in order to quiet the intense reaction. Addressing the underlying cause may not be possible unless the the excessive amounts of porphyrins in the body are treated first or concurrently. The chemistry of porphyria is distinctly different from that of conditions such as mold toxicity, Lyme disease, and mast cell activation syndrome, however, and therefore its treatment may be at odds with the treatment of those conditions, making the treatment for porphyria in the presence of these other conditions intrinsically difficult.

I often struggle to find the correct balance of what to treat, in what order, and how. The most obvious example is that treatment of porphyria requires a higher carbohydrate intake. To treat mold or *Candida*, we typically limit daily carbohydrate intake to 60 grams or less. It is fascinating that some patients find they feel awful until they get their carbohydrate intake up to 80 to 100 grams per day, and I wonder if these patients have a mild degree of porphyria. I have not studied this specifically, but I think it may be worthy of research.

An acute flare-up of porphyria, which looks to me like a prolonged Herx reaction (lasting from one to three weeks or more), may respond well to the use of intravenous (IV) dextrose in a 10 percent concentration (D10W), using 500 cc daily. In my experience, this intervention is the single most useful modality in settling down a flare-up. Often, one or a few IV treatments will produce a significant and rapid improvement in symptoms, greatly shortening the usual length of a porphyric reaction. It is not always easy for patients to obtain IV treatments, however. Instead, you can try taking dextrose (glucose) tablets to see if the reaction can be aborted. We have learned that an acute porphyrin attack stimulates the stomach to make a substance called PGC-1, which in turn stimulates the liver to increase heme production, worsening the porphyric episode. The use of intravenous 10 percent dextrose stops the production of heme and can ameliorate those attacks.

Because the treatment for mold toxicity and Lyme disease or *Bartonella* usually involves limiting carbohydrates and the treatment for porphyria is to markedly increase carbohydrate intake, you can see that these approaches are diametrically opposed and difficult to coordinate. Orchestrating treatment requires careful clinical judgment and usually involves increasing carbohydrate intake when a flare-up of porphyrins is suspected and resuming a low-carbohydrate approach when that flare-up has settled down. At the present time, from my perspective, this requires excellent guesswork and is inherently problematic.

Another simple treatment to quiet an attack of porphyria is to increase the level of carbon dioxide in the blood. There is good evidence that a low carbon dioxide level *(hypocapnia)* prevents the body from utilizing oxygen properly by depleting glutathione, which then depletes 2,3-DPG (diphosphoglycerate) in the red blood cells. This appears to be reversible, to a certain extent, by using rebreathing techniques such as breathing into a paper bag or by using the shallow breathing techniques developed by Dr. Konstantin Buteyko. The book *The Oxygen Advantage* by Patrick McKeown goes into these techniques in easily understood detail.

Low sodium levels, or *hyponatremia,* are common with porphyria, and taking a small amount of salt under the tongue may help with an attack. However, in some patients, salt can trigger an attack, so this must be approached cautiously.

Because the function of liver enzymes is essential to improving porphyria, supplements that assist this process, such as milk thistle and charcoal, which can adsorb excess porphyrins, may be helpful. In his book *Porphyria,* Dr. Rochlitz sets out a list of supplements and therapies that are of potential benefit to patients.

Another component of the difficulty of treating porphyria is the biochemical individuality that each patient brings to the table. The list of possible substances that have been shown to trigger porphyric reactions is long, and it is important for patients to realize that they may react to only a few. You will have to separately evaluate how you personally react to specific materials to learn how to work with this type of sensitivity. Unfortunately, gaining this information requires a great deal of trial and error, with the constant threat that if you stumble upon a triggering substance, you may be ill for days or weeks afterward.

> **In my experience, the most common trigger for porphyria is the treatment of *Chlamydia pneumonia* with antibiotics.**

In my experience, the most common trigger for porphyria is the treatment of *Chlamydia pneumonia* with antibiotics. Charles Stratton, MD, an infectious disease specialist at the Vanderbilt University School of Medicine, has discovered that killing *C. pneumonia* with antibiotics or the supplement NAC (N-acetyl cysteine) uniquely triggers a not-uncommon porphyric reaction. Because the antibiotics used for this purpose are so similar to those used for the treatment of *Bartonella,* they can inadvertently trigger a porphyric reaction in susceptible patients by releasing porphyrins from the killed *C. pneumonia,* causing what appears at first glance to be a severe and prolonged Herx reaction. Dr. Stratton has an excellent educational website, www.cpnhelp.org, that reviews this subject in detail and provides suggestions for treatment that are similar to what I outlined above. Dr. Stratton also notes that patients with secondary porphyria may require additional vitamin B12 and folic acid to both treat the porphyria and prevent recurrences.

I think of the interventions I have described as fairly basic, and I encourage patients who have porphyria to try them to see what works to quiet their inflamed systems.

Parenthetically, since resurrecting my interest in porphyria, I have checked twenty of my most sensitive patients with a LabCorp urine porphyrin test, and fifteen of them came back positive. This would seem to confirm the assertion that porphyria may be far more common in patients with marked sensitivities triggered by chronic inflammatory conditions than we had realized. Admittedly, this is a small sample, but I hope that other physicians and researchers will start looking at this issue as well.

The field of medicine that covers porphyria is endocrinology. If you suspect porphyria and hope to get assistance, please understand that endocrinologists view porphyria almost solely from the perspective of the rare genetic types. As such, unless lab testing shows levels of porphyrins that are ten times over the upper limits of normal, an endocrinologist is not likely to accept that diagnosis. With few exceptions, an endocrinologist will not consider the possibility of the secondary porphyrias that are the subject of this chapter. So, if you are hoping for help from a medical specialist, I fear you will be disappointed. It is up to those of us who care for these kinds of patients to get to the bottom of the causes of their illnesses and provide the best care we can. Until the medical profession embraces the understanding of environmentally acquired chronic inflammatory illnesses, we are on our own.

I am still attempting to integrate treatment for porphyria into my treatment strategies for mold toxicity, Lyme disease, *Bartonella,* and mast cell activation, and I find it to be quite a challenge. Each patient needs to be treated individually, and I don't have any clear "rules" or any kind of algorithm to follow. These patients tend to be extremely sensitive and toxic, so each treatment modality needs to be assessed separately so we can be assured that it is safe and helpful before moving on to the next component of treatment.

Carbon Monoxide Poisoning

Carbon monoxide has been shown to be a major cause of accidental poisoning for more than a century. It is produced by the incomplete combustion of any fuel; poorly vented gas stoves, gas or oil furnaces, space heaters, fireplaces, and cars running in closed garages are possible sources. Because it is an odorless, tasteless, and invisible gas, it can be difficult to know if you are being exposed to it. The use of carbon monoxide detectors or monitors can be lifesaving.

Acute toxicity can be relatively obvious, as in the case of finding a person lying unconscious (or worse) on the floor of a garage with the car's motor running. Chronic exposure is much less obvious, and discovering this possibility sometimes requires detective work.

A review of the symptoms of chronic carbon monoxide poisoning reads exactly like the symptoms we find for porphyria, mold toxicity, Lyme disease, and fibromyalgia. A multisystem listing would include:

- Fatigue
- Weakness
- Nausea
- Confusion
- Dizziness
- Chest pain
- Headache

Especially important to this discussion is that carbon monoxide also amplifies a person's sensitivity to light, touch, sound, smells, and tastes. If you are interested in exploring this subject in more detail, I encourage you to read the article "Background on Sources, Symptoms, Biomarkers and Treatment of Chronic Carbon Monoxide Poisoning," by Albert Donnay, MHS, available at www.mcsrr.org.

Making the diagnosis of chronic carbon monoxide poisoning can be difficult. When a patient presents to an emergency room suspecting exposure, the standard protocol is to test the patient's carboxyhemoglobin level. If done several days later rather than immediately upon exposure, this test becomes inaccurate. Another test that can point to this diagnosis is a blood test for venous blood gases, in which elevated levels of oxygen in venous blood (called PvO2) might be found. Measuring carbon monoxide in exhaled breath also can be used to make a diagnosis, but this measurement is not readily available.

Complicating the possibility of exposure to carbon monoxide is that the body is capable of making carbon monoxide under stress. Stress can stimulate the increased production of an enzyme called heme oxygenase-1 (HO-1), which breaks down heme from heme proteins, creating a variety of metabolites that include carbon monoxide. You can see how this relates to porphyria, in which the breakdown of heme is similarly disordered.

The take-home message here is that in a patient who has become extremely sensitive to everything—light, touch, sound, smells, and tastes—with its accompanying hypervigilance and anxiety, the possibility of carbon monoxide poisoning should be considered as the source.

The treatment for this condition primarily utilizes 100 percent oxygen inhaled through a nasal canula from a tank or oxygen concentrator at a rate of 6 to 10 liters per minute for two hours a day. Healing often requires three to four months of treatment. Unfortunately, it can be difficult to obtain oxygen for home use because the medical prescription of oxygen requires that specific criteria be met to allow distributors of oxygen devices to do so. Patients who might benefit often do not meet these requirements.

Albert Donnay did a study of thirty-four patients with ME/CFS and noted improvement in all patients treated with oxygen in this way, with 25 percent reporting marked improvement in all sensitivities and 38 percent reporting improvement in their chemical sensitivities.

William Rea, MD, a pioneer in the treatment of patients with multiple chemical sensitivities, has found the daily use of oxygen in his clinic to be of great benefit. I have seen a number of patients with extreme sensitivities improve with oxygen treaments as well. The possibility that chronic carbon monoxide toxicity plays a role in the development or exacerbation of intense sensitivities clearly needs to be explored in greater depth.

What I hope I have conveyed in this short chapter is the need to consider porphyria and/or carbon monoxide poisoning as a possible contributing factor in the evaluation of sensitive and toxic patients. Treating it may be of great benefit to quiet some patients' overly excited nervous systems so that healing can move forward.

" In a patient who has become extremely sensitive to everything— light, touch, sound, smells, and tastes— with its accompanying hypervigilance and anxiety, the possibility of carbon monoxide poisoning should be considered. "

CHAPTER 7

Where to Start?

Now that we have reviewed what are, in my experience, the most common causes of increased sensitivity and toxicity, we should have a starting point for treatment. If mold toxicity, *Bartonella* infection, mast cell activation syndrome, porphyria, carbon monoxide poisoning, or some combination thereof is present, treatment should begin there. The chapters in this part of the book outline specifics for how to begin treatment for each of those conditions, and I believe that these initial steps should be the underpinning of all treatment efforts.

One of the more obvious questions, then, is: *Given that there are four major conditions that can incite this kind of extreme sensitivity and toxicity, where should we start treatment?*

It should come as no surprise that the answer is: *It depends on the patient.*

If mold toxicity emerges as the priority, that can be the starting point, but if the patient has mast cell activation or porphyria, those conditions may need to be quieted down before the patient can tolerate more specific treatment for mold or *Bartonella*. Sensitive patients often cannot even begin to tolerate treatment until their nervous systems have settled down. As you will see in Chapter 9, efforts to quiet down a patient's overreactive nervous system might include the use of Annie Hopper's Dynamic Neural Retraining System (see page 156) or the addressing of the autonomic nervous system using our new understanding of Polyvagal Theory. It might require the use of tiny doses of supplements that can ease the patient's body into beginning to be able to detoxify. If mast cell activation is in play, addressing it early in the course of treatment allows the patient to become much less reactive and much more capable of handling the rest of the treatment program.

The same is true for porphyria, if it is present. In an unusually sensitive patient, therefore, we should look for mast cell activation and/or porphyria and, if either is likely to be present, start by treating that condition.

When both mold toxicity and *Bartonella* are suspected in a sensitive patient, we start with the mold, with few exceptions. There are several reasons for this approach:

- Treatment for mold is gentler and better tolerated overall.

- Treatment for mold does not require the extensive use of antibiotics, which are necessary for most patients with *Bartonella*, thereby sparing their delicate gut microbiomes in the early stages of treatment.

- Mold toxicity and *Bartonella* infection are remarkably similar in their presenting symptoms. Once the mold has been successfully removed, the symptoms attributed to possible *Bartonella* may have disappeared entirely, meaning that *Bartonella* no longer needs to be considered as an issue requiring treatment.

I have come to view a treatment program as being orchestrated in layers. The outermost layer is the one in most need of attention. It presents itself by the symptoms that are most prominent in a patient. If, after the treatment for mold toxicity, mast cell activation, and/or porphyria is completed, the patient is better but still has symptoms that have not fully resolved, we look for the next layer, which will convey itself via changes in those symptoms. By now, the patient should be nowhere near as sensitive as he or she was at the beginning of treatment and should be able to tolerate the next mode of treatment far more easily.

I urge healthcare providers not to be distracted by other findings on lab reports that would lead their treatment efforts elsewhere. If you start treatment based on the possible presence of viral components or methylation imbalances or genetic possibilities as discovered by measuring SNPs (see Chapter 16) or heavy metal toxicities or whatever else it may be, I suspect you are missing the point. Keep in mind, though, that this book is devoted to the evaluation and treatment of unusually sensitive and toxic patients, so a patient with a healthy constitution may well be able to tolerate and benefit from much more extensive treatment.

PART 3

REBOOTING AS A MODEL FOR TREATMENT AND HEALING

If the Shoe Fits...

In his work "Antipurinergic Therapy for Autism—An In-Depth Review," Dr. Robert Naviaux emphasizes that "in chronic illness the original triggering event is often remote and may no longer be present." It is imperative that we remember this as we evaluate and treat chronically ill patients.

Our new understanding of chronic illness as having been created by a persistent inflammatory response to a stimulus that may no longer be present changes the entire paradigm of how we look at diagnosis and treatment. Our first job, as outlined in Part 2, is to look for conditions that are keeping the inflammatory process going, such as toxins, infections, and other biochemical disturbances. Once we have successfully treated those, we are increasingly finding that patients are still not well. Somehow, that individual's body has lost the ability or knowledge of how to turn off the process that started the illness. So we need to "reboot the system."

For those who know me, the use of a rebooting metaphor will be highly amusing. When it comes to technology, admittedly I am a dinosaur. My knowledge of and use of computers is limited. I can answer emails, use word processing programs, and make Power-Point presentations, and that's about it. If I cannot figure out how to do something on my computer, my long-suffering wife will roll her eyes and agree to help me once again. When my youngest son left home, I was lost. Despite this, it is not lost upon me that when my computer locks up, I need to power it down and start it up again. I suspect that almost everyone has had that same experience. When my computer locks up, it simply does not work. This highly refined machine is now useless to me. And, of course, this lockup normal-ly occurs when I am pressed for time and need some computer function to work *now*. This is the best metaphor I have come across for what happens to the body when it gets "stuck." It locks up and stops working properly.

This is what happens with the illnesses that were discussed in Part 2. Even when the cause(s) for getting locked up or stuck have been addressed, nothing will work until the "computer" is rebooted. If we turn it off and then turn it back on, it magically starts working again.

We are increasingly discovering that in patients who have been ill for long periods, their systems do not always "reboot" auto-matically once the problems have been corrected. And for those patients and their healthcare providers, I am including this infor-mation that looks not only at the need for rebooting, but also at all of the methods I have learned about and utilized to reboot these biological systems and restore them to normal. This will not be a comprehensive list but will focus on the pieces I have found to be of greatest clinical value. I hope that this discussion provides a lot of hope and direction for patients and healthcare providers alike in the move toward complete healing.

First, we will explore the groundbreaking new model for chronic illness developed by Dr. Robert Naviaux, called the *cell danger response,* in Chapter 8. Then we will look at each of the major systems of the body that get "stuck" and explore the methods of rebooting those systems that have worked for my patients, system by system. Chapter 9 begins with rebooting the nervous system, because that is the easiest to understand as a rebooting model. We will turn our attention to the immune system in Chapter 10, as inflammation is a key component of many chronic illnesses, and then move on to the endocrine system in Chapter 11. Chapters 12 and 13 discuss rebooting the gastrointestinal system and addressing the associated weight gain that is so problematic for many patients. Chapter 14 covers methylation dysfunction, which is almost universally present in my patients. Chapter 15 is devoted to detoxification and how to bring it into the healing process, while Chapter 16 explores the genetic conditions that predispose certain people to chronic illness. Chapter 17 delves into multiple chemical sensitivies (MCS) in more detail. Finally, Chapters 18 to 20 shift gears to address the important realms of mental, spiritual, and emotional blockages that are critical to engendering the healing process.

For some of these chapters, I have asked colleagues to chime in with specific information and stories of their patients to broaden the scope of our discussion. My intent in doing so is to allow this to be less "my" book and more "our" book.

CHAPTER 8

Rebooting as a Model for Treatment: The Cell Danger Response

This chapter is central to the understanding of how cells get "stuck" in a reaction that was intended to heal the body but, when prolonged, backfires and becomes part of the problem instead. I have tried to simplify some of this information, but it is complicated biochemistry and physiology. Although this book is intended for both physicians and patients, I suspect that this chapter will be more readily appreciated by physicians. Still, many of my patients are extremely well read, and it is my hope that many of you who are suffering from these sorts of conditions will be able to work your way through this chapter and will find this information to be of great value. If it seems overwhelming, please just skim it and move on.

I have discussed a wide array of conditions that can trigger an out-of-control inflammatory process. Dr. Ritchie Shoemaker's brilliant Biotoxin Pathway model (outlined in Chapter 3) helped us put together some of this complexity by showing how a toxin can trigger a choreographed sequence of reactions leading inexorably, like a stack of biochemical dominoes, to inflammation, immune dysregulation, hormone dysregulation, gastrointestinal imbalances, pain, insomnia, fatigue, and weight gain. In short, we have an explanation for how so many different symptoms can be caused by one triggering entity.

Dr. Shoemaker later expanded his model to include the concept of Chronic Inflammatory Response Syndrome, or CIRS. As we began to see the inflammatory underpinning of chronic fatigue syndrome, fibromyalgia, mold toxicity, Lyme disease, chronic viral infections, neurodegenerative conditions, PANDAS, PANS, and autism spectrum disorders, we began to realize that there is a common denominator in all of these conditions—that given an individual's unique biochemistry, age, and genetic predispositions, the same insult to the body could create a wide variety of symptoms. At their essence, these conditions are more similar than we had realized.

In 2013, Dr. Robert Naviaux, professor of genetics in the Departments of Medicine, Pediatrics, and Pathology and co-director of the Mitochondrial and Metabolic Disease Center at the University of California, San Diego, School of Medicine, published his ground-breaking model of how this information can be understood on a deeper biochemical level. He calls it the *cell danger response.*

Dr. Naviaux looked at the complicated biochemical abnormalities that are seen in these inflammatory conditions and recognized that they all could be explained as a deeply ingrained, primitive response to a cell being threatened by an infectious agent or toxin. Using a virus entering a cell as an example, Dr. Naviaux outlines a clear sequence of events to clarify how that cell is designed to respond to such threats. This is the cell danger response, or CDR. While a complete description of this process is beyond the scope of this book, I would like to try to simplify the CDR so that we can apply its principles to improving our ability to diagnose and treat it.

So here goes: Using the example of a virus entering a cell, the first thing that occurs is that the mitochondria in the cell (the organelles that we rely on to make energy) detect the presence of that intruder as a diversion of electrons, or a voltage drop—a decrease in electron flow within the mitochondria. As soon as they sense this drop, the mitochondria react—almost instantaneously (in a process that has evolved over centuries)—to decrease their oxygen consumption. This might seem odd; wouldn't the cell want to rev itself up to deal with the threat? But what happens is that the oxygen concentration within the mitochondria rises, making cellular redox chemistry more oxidizing, *to shield the cell from further injury.*

It is very important to understand that this is a protective mechanism. I apologize for the technical language that I am using to describe it; it is not necessary for you to completely understand this complicated series of events. But we must recognize that this defense mechanism has been evolving since the beginning of time and that it is deeply ingrained in our cells. This is how we protect ourselves from threats.

This rise in mitochondrial oxygen concentration triggers eight additional events as part of this protective mechanism (see Figure 8.1):

1. Cellular metabolism shifts to prevent the hijacking and assembly of cellular resources by intracellular pathogens. The invading virus, for example, is just a protein coating around an RNA center. It cannot make what it needs for itself; it has to take advantage of the host's resources to make those materials. So our cells have learned to attempt to deny the invading virus what it requires to grow and thrive, thus preventing it from replicating.

2. The cell membrane of the invaded cell stiffens to prevent the virus (or other pathogen) from leaving the cell. This might seem to you like the cell is committing hara-kiri—and it is. The cell essentially sacrifices itself for the greater good with the hope that the virus will be unable to spread to other cells.

3. The cell releases antiviral and antimicrobial chemicals into the area around the cell to further limit the spread of the infectious agent.

4. Autophagy and mitochondrial fission increase to remove intracellular pathogens. To explain this language, *auto* simply means "self" and *phagy* means "to eat," so the cell releases chemicals to destroy itself in the hope of destroying the invader as well. (Bob Miller discusses autophagy in more detail in Chapter 16.)

5. DNA methylation and histones are changed to alter gene expression. The invader cannot methylate. Methylation (discussed in Chapter 14) is required to make or repair DNA, and a virus can't replicate its DNA or RNA (translation: grow) without hijacking our methylation chemistry. So we shut down our own methylation chemistry to limit the virus.

6. Endogenous retroviruses and other genetic elements are mobilized to produce genetic variants; in other words, we attempt to change our genetics to make ourselves less susceptible to the attacker.

7. Neighboring and distant cells are warned of the danger. The invaded cell sends out chemical messengers that ring the alarm: DANGER! DANGER!

8. The behavior of the host is altered to limit the spread of infection: fatigue forces us to rest, and we sleep more to conserve energy.

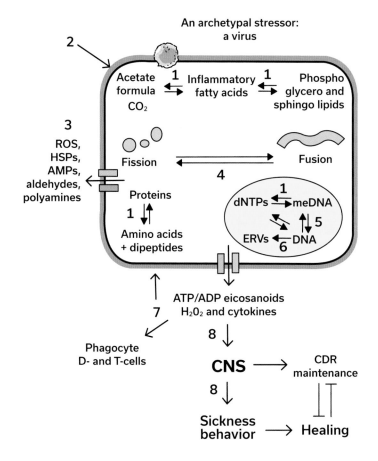

An archetypal stressor: a virus

2

3
ROS,
HSPs,
AMPs,
aldehydes,
polyamines

Acetate formula CO_2 — 1 — Inflammatory fatty acids — 1 — Phospho glycero and sphingo lipids

Fission — 4 — Fusion

Proteins 1

Amino acids + dipeptides

dNTPs — 1 — meDNA
ERVs — DNA
6 — 5

ATP/ADP eicosanoids H_2O_2 and cytokines

7

Phagocyte D- and T-cells

8

CNS → CDR maintenance

8

Sickness behavior → **Healing**

Electron Steal Drives Rapid Mitochondrial Redox Changes

0. Decrease oxygen consumption → increase dissolved O_2 concentration

1. Shift from polymer to monomer synthesis

2. Stiffen cell membranes

3. Release antiviral and antimicrobial chemicals

4. Increase mitochondrial fission and autophagy

5. Change DNA methylation

6. Mobilize endogenous retroviruses and LINEs

7. Warn neighboring cells and call in effector cells—the "purinergic halo"

8. Alter host behavior to prevent spread of disease to kin

Figure 8.1. A systems view of the cell danger response (CDR).

The next step is to understand some of the vital chemical reactions in the body that are affected by these protective events. Grasping the major imbalances that are triggered by an invading microbe or toxin gives us a blueprint for how to go about repairing them. Though complicated, we must understand this chemistry in order to help orchestrate the healing process.

Before I launch into the chemistry, I would like to discuss another important component of the CDR: the shift (or inability to shift) from summer metabolism to winter metabolism.

While this subject has not received much attention, over the millennia humans have evolved two different systems of metabolism. From our ancient roots, when food was plentiful in summer, our ancestors fed well on a variety of foods and gained weight and flourished. In winter, when food was scarce and variety was limited, people ate much less; this was a time when metabolism regrouped. On a cellular level, this dormant period was used to rebuild and repair the body's internal structure. Metaphorically, you might think of this as winter on the frontier, with a pioneer family sitting around the stove for hours, staying warm, mending and repairing what needed to be done in preparation for the arrival of spring.

It turns out that we have evolved two different biochemistries, one based on how we metabolize food in the summer and the other based on how we metabolize food in the winter. These two biochemistries are orchestrated by two different modulators. *mTOR* (mammalian target of rapamycin) is the master fuel sensor in the cells during the summer months. mTOR facilitates protein synthesis and growth by using new nutritional materials taken in from the environment. This promotes rapid growth without inflammation. In winter, the master fuel sensor switches to *AMPK* (AMP-activated protein kinase), which optimizes energy efficiency and stimulates the recycling of cellular materials for repair.

Figure 8.2, put together by Dr. Naviaux, summarizes some of the key biochemical shifts between summer and winter metabolism. These shifts are seasonal, but they also occur to a lesser extent every night during sleep and can be stimulated by fasting. The biochemical pathways stimulated by AMPK in winter metabolism support regeneration and are anti-inflammatory, as they work to break down damaged proteins, lipids, DNA, and RNA.

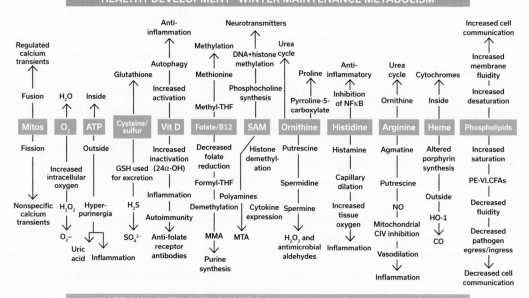

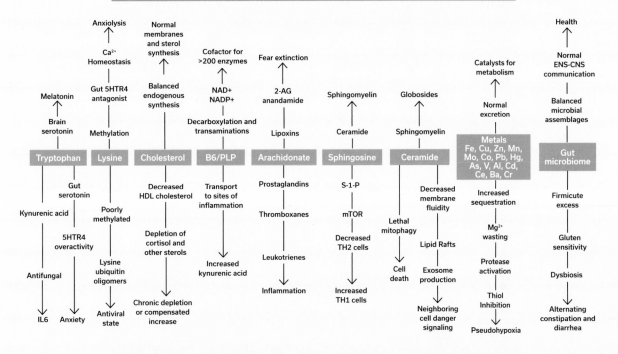

Without getting caught up in the details, what this means is that if we cannot cycle naturally between summer and winter metabolism as our bodies are designed to do, we can get stuck. As you might guess, by living in a metabolic state that is mostly "endless summer"—where we can eat any food we like in any season, without restriction—we are losing our natural movement toward (and denying our natural need for) winter metabolism. Although there are many explanations for the profound increase in obesity, one explanation that does not receive much airtime is the inability to shift into winter metabolism.

For patients in whom inflammation is a key player, the distinction between summer and winter metabolism takes on added importance. The kinds of illnesses we have been discussing—chronic fatigue syndrome, fibromyalgia, Lyme disease, mold toxicity, and chronic viral infections—all share persistent inflammation that the body appears powerless to control. If we could re-establish the natural balance between summer and winter metabolism, is it possible that we would gain another approach to shutting off these relentless inflammatory signals?

Encouraging fasting could stimulate the transition to winter metabolism. While many patients are put off by the idea of fasting or its presumed difficulties, simpler, less-challenging variations are emerging. Simply skipping dinner provides what is called a "modified fast" of sixteen hours. For some patients, a fast of this duration would be sufficient to facilitate the shift. Others could handle longer fasts under medical supervision. For those with diabetes or hypoglycemia, however, fasting can present difficulties, so fasting strategies should be undertaken only under medical supervision.

Specific and Potentially Treatable Features of the CDR

As technical as it still may seem, I have gone to some effort to simplify a sophisticated biochemical discussion. Please don't get caught up the complexity of the chemistry: *what is really important is that we restore biochemical balance in order to turn off the CDR alarm.* Turning off that alarm may be the single most important thing we do to move a patient toward the possibility of healing. It also is vital to keep in mind that turning off the CDR alarm before we have appropriately dealt with whatever threat has set off that alarm is not a good strategy. The following are the eight major biochemical shifts delineated by the CDR:

> **NOTE**
>
> To understand the CDR, some knowledge of biochemistry is essential. For some readers, the following paragraphs may be difficult to grasp, and it is not necessary that they be understood. If this material is intimidating to you, please move on to the next section. This is the most biochemically detailed description in this book.

1. Stressed cells release purinergic signaling molecules, like ATP. These molecules activate the process of inflammation (technically by activating the NLRP3 inflammasome assembly), and this, in turn, stimulates the body to make more cortisol. Of particular interest here is that this process is independent of ACTH (adrenocorticotrophic hormone) production—meaning that it is separate from the pituitary gland's attempt to regulate stress hormones. This process is set off directly by the CDR. *Of all the features listed here, this one may be the most immediately amenable to treatment and is critical in the rebooting of the CDR.* Dr. Naviaux recently published a groundbreaking preliminary study of the benefits demonstrated by the use of suramin, a purinergic blocking agent, in the treatment of autism (see below).

2. Sulfur metabolism shifts so that glutathione (an essential material for detoxification, energy, and communication) is consumed in an attempt to regulate detoxification. Cysteine, which is ordinarily utilized to make more glutathione, is diverted to making hydrogen sulfide and taurine. This process interferes with methylation, a key component for restoring chemical balance. (See Chapter 14.)

3. Vitamin D metabolism is altered. You may think of vitamin D as mostly about bone health, but it is much, much more than that. Vitamin D plays a critical role in dealing with inflammation and preventing autoimmunity, in which a confused immune system attacks parts of itself. So threats of danger, which may come from exotoxins, may increase the activity of the enzyme 24-alpha hydroxylase, which in turn decreases the concentration of active vitamin D and directly increases inflammation and the risk of autoimmunity.

4. The CDR directly stimulates the enzyme histidine decarboxy-lase (which is dependent on vitamin B6) to produce histamine, which is critical for mast cell and eosinophil cell functioning in their coordination of allergy and antiparasitic immunity. (Refer to Chapter 5 for a more detailed discussion of mast cells.)

5. When the CDR is activated, red blood cell and mitochondrial heme centers are released from damaged cells, interfering with porphyrin synthesis. (Refer to Chapter 6 for a more detailed discussion of porphyrins.)

6. The metabolism of tryptophan, an essential amino acid, may be altered in several ways in response to the CDR. By the hydroxylation pathway, it may be converted to serotonin or melatonin. By the dioxygenase pathway, it may be converted to kynurenic acid or quinolinic acid, which in turn induces more inflammation. You may not have heard of kynurenic or quinolinic acid, but elevated levels of these acids have been associated with intense anxiety, depression, and other psychological issues, so for many patients, this is indeed important.

7. The effects of the antiviral CDR are strongly regulated by the effects of the amino acid lysine. It turns out that dietary lysine is an antagonist of gut serotonin receptor 4 and accordingly has anti-anxiety effects and opposes the CDR. Taking oral lysine, therefore, emerges as a strategy for rebooting the CDR.

8. The CDR produces low plasma levels of pyridoxal 5-phosphate (P5P), the important active form of vitamin B6. (Pyridoxine is another name for vitamin B6.) Of particular note here is that a deficiency in P5P shifts the metabolism of tryptophan (noted in point #6 above) toward kynurenic acid, which further promotes the inflammatory process, maintaining the CDR.

9. Under healthy conditions of metabolism, toxic metals do not accumulate because our ability to excrete them normally exceeds our exposure to them. When the CDR is activated, however, the internal milieu of the cell is changed such that it favors the sequestration of toxic metals. Toxic amounts of trace and heavy metals thus may accumulate and may not be easily excreted. Translation: Patients may present with significant heavy metal toxicity that they have difficulty getting rid of, which adds to their overall toxicity and prolongs their illness.

10. When the host is sick, the gut microbiome also is sick. When chronically activated, the CDR alters the physical habitat of the

bowel with profound shifts in the biome. What this means is that virtually all patients who are chronically ill have "leaky gut" from chronic inflammation and dysbiosis, which leads to an increase in food allergies and the possibility that other pathogens, such as *Candida*, pathogenic bacteria, and parasites, will become more prominent. (The gastrointestinal system is discussed in more detail in Chapter 12.)

11. The CDR is initially *adaptive* and coordinated by the close interplay of the mitochondria and the cell. However, as it persists, it may become *maladaptive* once the initial danger is gone. If the cell cannot turn off the alarm that it has set off, the CDR becomes part of the problem rather than the solution.

Phew. You made it; you got through that section. I will get back to writing English again.

Of particular importance is the following statement that Dr. Naviaux makes in a paper published in 2017, entitled "Antipurinergic Therapy for Autism—An In-Depth Review":

In chronic illness the original triggering event is often remote and may no longer be present.

Let me restate that: Once the danger has passed, unless the alarm can be turned off, this process will have a life of its own and will persist. This may seem like a simple statement, but there is a lot to dissect here. First, let me emphasize the "Once the danger has passed" part. If the body is dealing with an infection and/or a toxin, if either or both of those remains untreated, turning off the alarm will be exceedingly difficult. Well-meaning attempts to begin rebooting the system or to tell the cell that it is no longer in danger will go nowhere if the cell knows differently. To put this another way, if mold toxicity or Lyme disease or other infections remain undiagnosed or have not been treated successfully and are still present, attempts to turn off the CDR are unlikely to be effective. This means that identifying and treating the causes for the CDR must be paramount in our thinking. Only after that has been achieved can we proceed with rebooting strategies to reverse this process and get the patient moving toward healing. Using the blueprint provided by Dr. Naviaux's understanding of the CDR gives us a clear method for how to proceed.

I have tried (believe it or not) to simplify this model and make it as understandable as possible for the readers of this book. To continue this simplification, taking the information already outlined, we can now discuss how to use that model for treatment.

Since the publication of the CDR in 2013, many health professionals from a wide variety of disciplines have hailed this model as a breakthrough in our understanding of chronic illness. To increase our understanding of this process, in the 2017 paper just cited, Dr. Naviaux describes two different types of mitochondria, designated as M1 and M2. Another way of describing the CDR using these designations is that the CDR shifts mitochondrial function so that the M2 mitochondria, which perform all those wonderful energy-creating functions when we are healthy, are converted to M1 function. The M1 mitochondria are specialized to create the oxidative shielding response—triggered by the CDR—and perform dozens of antiviral and antimicrobial functions in the service of healing. Changing this response would require a shift from M1 to the more normal M2 mitochondria, but sometimes the cell does not realize it is ready for that shift and remains "stuck" in M1.

Using the CDR as a Model for Diagnosis

For a study outlined in his groundbreaking paper "Metabolic Features of Chronic Fatigue Syndrome" (of which I am honored to be a co-author), Dr. Naviaux utilized liquid chromatography/mass spectrometry technology to measure 612 chemical substances in the blood of patients with chronic fatigue syndrome (CFS) and then compared them with age- and sex-matched controls. We tested twenty-two men and twenty-three women with CFS and an almost equal number of controls. Of the 612 substances that were examined, CFS patients consistently showed deficiencies in just a few. In fact, looking at just eight of these substances in men and thirteen in women enabled us, for the first time, to diagnose CFS with an accuracy of 94 percent in men and 96 percent in women! Granted, these findings will need to be confirmed in larger studies (one of which has been completed and will be published shortly), but it is clear that CFS cannot be thought of as a psychological illness when measuring just a few substances with a single blood test tells us with great accuracy who does and does not have it.

Several other findings from our research are important. First, the biochemical profile of these deficiencies is different in men than it is in women. While there is some overlap, this confirms the importance of appreciating the biochemical differences between the sexes. We have clear confirmation that men are indeed from Mars and women from Venus. (See Figure 8.4.)

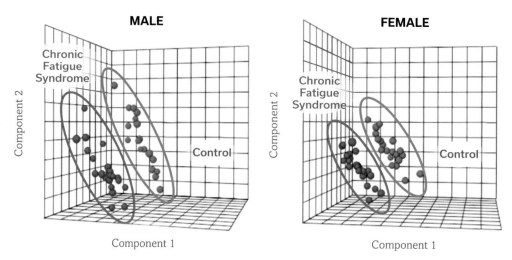

Figure 8.4. CFS patients show deficiencies in certain substances, which have different biochemical profiles in men and women.

Second, while all of the subjects had certain key deficiencies, there were significant differences in *which* substances were deficient or elevated for 75 percent of these substances. In other words, 25 percent of the deficiencies we discovered were shared by all patients, but the other 75 percent differed from patient to patient. This confirms the often-stated premise that we must look at patients individually to understand how to treat them. *Even though all of them meet the classic definition of CFS, it affects them differently.* Medicine has longed for a simple algorithm for treatment, and it is now clear that this algorithm is not going to materialize.

The biochemical substances in which the majority of our CFS patients were found to be deficient are fascinating, even if the names of these substances are not household words. (I suspect that they will be in a few years.) They include abnormalities in the metabolic pathways of phospholipids, sphingolipids, purines, riboflavin, P5P, branched-chain amino acids, cholesterol, microbiome, peroxisomal, and biochemical measures of mitochondrial function (see Figures 8.5 and 8.6). While obviously complicated, all of these changes can be understood as likely representing an adaptive cellular response designed to oppose the spread of intracellular bacterial and viral infections.

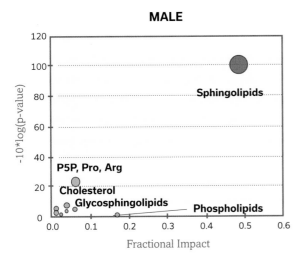

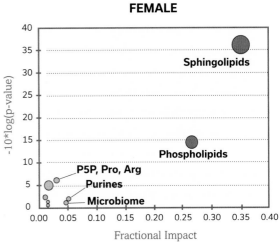

Figures 8.5 and 8.6. The biochemical substances in which CFS patients were found to be deficient.

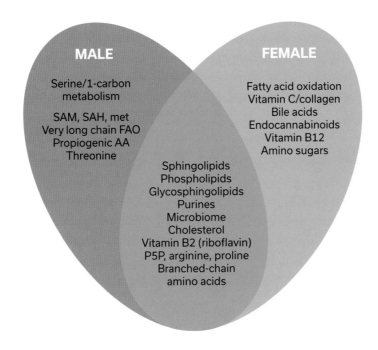

MALE

Serine/1-carbon metabolism

SAM, SAH, met
Very long chain FAO
Propiogenic AA
Threonine

FEMALE

Fatty acid oxidation
Vitamin C/collagen
Bile acids
Endocannabinoids
Vitamin B12
Amino sugars

Sphingolipids
Phospholipids
Glycosphingolipids
Purines
Microbiome
Cholesterol
Vitamin B2 (riboflavin)
P5P, arginine, proline
Branched-chain
amino acids

• • • •

There are three sharp raps on my door, and a high-pitched male voice demands insistently, "Neil!" This is followed by three more sharp raps and a repeat of my name, "Neil!" And again, once more, "Neil!"

I open the door and say, "Yes, Sheldon, what is it?"

"I just wanted to be certain your readers realize that this complicated biochemistry you are attempting to explain will be the future of medicine in just a few years."

"Ah," I say, "and you felt so strongly that I had not done so properly that you felt compelled to disturb me tonight to let me know?"

Sheldon just rolls his eyes and allows the door to close gently behind him.

Thank you, Sheldon.

• • • •

But Sheldon is right. I confess that only a few physicians will immediately understand this chemistry, so some of what I have just explained must sound like gibberish to most of you. And maybe, just a little, I am showing off that I can use some of these words and abbreviations in a sentence. My intention, however, was simply to introduce you to information and biochemical terms that I anticipate will become commonplace in a few years. The testing used in this research is not yet available commercially but will be soon. In fact, very similar technology is now being used by the Great Plains Laboratory for its new mycotoxin test, which was discussed in Chapter 3.

Dr. Naviaux's lab is pioneering the use of metabolomics testing, this process we have been discussing in which 612 different biochemical substances can be measured in a single specimen of blood or urine, and I believe that this study will go a long way toward changing the ways in which we evaluate and diagnose patients with chronic illness. Our old standby, the CMP (comprehensive metabolic profile), will be viewed as inadequate for providing the information needed. How many patients have had dozens of CMP tests performed, all of which were ostensibly "normal," when it is clear that they are still sick? As Einstein suggested, repeating the same test and obtaining the same negative results, with no resulting clarity or benefit, makes no sense whatsoever. We now have a tool that can do much better.

Using the CDR as a Model for Treatment

Here is where this complicated information becomes immediately useful.

Because methylation chemistry is altered as part of the CDR, treating methylation (when *timed properly,* as discussed in Chapter 14) is an obvious component of treatment. So is addressing vitamin D intake following the measurement of vitamin D in the blood. So is the addition of lysine to the treatment program, if warranted. So is assessing and treating the gut microbiome (see Chapter 12). So is addressing the intake of vitamin B6 and the amino acid tryptophan. So is looking at histamine production (see Chapter 5) and heme metabolism (see Chapter 6). And these are only the basics to look at following the eight major biochemical shifts delineated by the CDR.

When we have the full metabolomics picture available to us, which I hope will be soon, we will have a more complete blueprint to work from that looks at a much more complete evaluation of biochemical deficiencies and excesses that will be *specific to each patient.* This will allow us to provide tailor-made treatment programs designed with the patient's unique needs front and center. While we can currently provide fairly decent educated guesses about what a patient needs, this evaluation will give us the scientific information that we require to deliver exactly what the patient needs, when he or she needs it. Understanding the CDR, therefore, has the potential to revolutionize how we practice medicine. I believe it is our future.

Congratulations! You have made it through the most difficult section of this book. The rest should be relatively easy.

CHAPTER 9

Rebooting the Nervous System
Turning Off and Resetting the Alarm

The most obvious body system to which the rebooting metaphor applies is the nervous system, which is intrinsically electrical in nature. When an electrical system is not functioning properly, messing with it often makes things worse. The computer analogy that I referred to in the introduction to this part is particularly apropos and is worth repeating. When a computer "locks up," we have learned that fixing it is simple: we turn off the computer, wait a minute, and then reboot it. And voilà, it works.

This process is similar to what happens in the human nervous system when things are not working properly. A variety of physiological events interfere with the precise signaling that the nervous system requires to function properly. The main one appears to be persistent inflammation, no matter what the cause. You can see how this applies to my patients: one of the primary components of chronic illness is persistent, uncontrolled inflammation. The swelling and anatomical distortion that accompany inflammation change the relationships between nerve endings, interfering with the electrical signaling that is the essence of the nervous system.

In fact, one of the best-accepted explanations for chronic pain is that of the *reverberating loop.* The idea is that when a pain signal comes into the spinal cord (from anywhere in the body), it first makes contact with connecting neurons in the dorsal horn of the spinal cord, which passes that electrical signal on to the brain, where it stimulates the proper receptor to let us know that we have run into something that is hurting us. (See Figure 9.1.) In the region of the spinal cord in which this connection is made, the nerves are sometimes packed together so densely that rather than moving directly to the brain, the signal stimulates neighboring nerves, which loop back around and repeat that signal to neighboring nerves, and *then* proceeds to the brain. This looping process helps us understand how pain can become chronic. The pain signal sets off a looping process such that the nerves in the spinal cord that pass the signal on to the brain continue to be stimulated, *even when the original pain signal is long gone.*

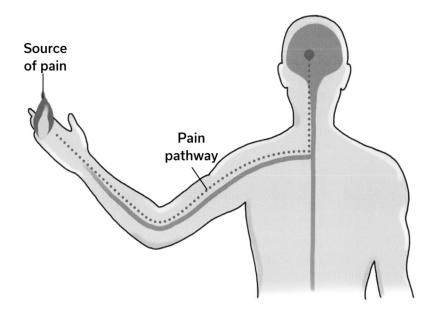

Source of pain

Pain pathway

Figure 9.1. The pain signal comes into the spinal cord and moves up the brain.

For example, say you overexert your lower back while lifting a heavy object, which triggers a pain signal. Your brain gets that signal and lets you know by clearly manifesting the message: "Oooow! You've got pain in your lower back!" If that signal triggers a reverberating loop, your brain will still think you are in pain even after the injured area heals, and it will tell you so very clearly. A key principle of chronic pain management is to find a way to break ("reboot") this reverberating loop so that the signal ceases and the brain finally realizes that the injury has healed. The use of acupuncture and a wide variety of electrical treatments (such as TENS, FSM, and injections) can be quite helpful for rebooting; these treatments are described later in this chapter.

The inflammatory processes that underlie chronic illness are predisposed to trigger this type of reverberating loop. Because mold toxicity, Lyme disease, and several of its co-infections, particularly *Bartonella,* have a propensity to inflame nerve tissues and the brain, they often set off a wide array of unusual neurological symptoms. As previously discussed, these include odd tics or spasms (which are medically termed *dyskinesias*), seizure-like tremors and spasms (which are medically termed *pseudoseizures* because, when observed by neurologists, they look like seizures, but repeated EEG testing does not demonstrate the typical brain wave patterns that are common to the diagnosis of seizure disorders), electrical sensations, shocking pains, paresthesias in areas of the body in which these symptoms are not usually noted, mood swings, severe anxiety and/or depression, and cognitive impairment.

Sometimes, even when the mold toxicity and/or infections have been properly treated, these neurological symptoms persist. This persistence can be quite confusing, adding to the difficulty of diagnosis because they imply that the infection or toxin still may be present. However, all too often these ongoing symptoms are a manifestation of this reverberating loop, meaning that somehow the neurological trigger is stimulating itself, and the body does not know how to turn it off. If this sounds similar to the cell danger response and an immune system that does not know how to shut off the signal that it is inflamed, as described in Chapter 8, it most certainly is.

I have seen many patients who were correctly treated for their illness but continued to be treated erroneously for years when their symptoms had not fully resolved, with the mistaken assumption that the illness was still active, because their physicians did not understand this concept. Conversely, I have seen many patients improve dramatically when we found ways to reboot their nervous systems, and it then became clear that the treatment for infection or toxicity was complete. As you might expect, discerning exactly where a patient is in this process can be tricky. Are we looking at a symptom of illness or of neurological "looping"? Very few tests are available to clarify the results of treatment in these controversial areas of medicine; simply attempting a rebooting process is often gratifying.

I have seen a wide variety of treatments accomplish this rebooting of the nervous system. In this chapter, I will discuss the ones with which I am the most familiar and the ones that have worked best for my patients. Please do not leave with the impression that these are the only known rebooting processes—far from it. I hope that you leave this chapter with a new optimism that many as-yet-untried treatments might be of great help to you. Some of these modalities are specifically rebooting processes, and others are biochemical imbalances that need to be rectified in order for the rebooting process to work. For example, if a patient's magnesium levels are low, even the perfect treatment will work briefly and then wear off; the patient's body won't be able to sustain it. Often in these cases, patient and physician get a little bit of hope and then abandon the treatment prematurely because they are under the mistaken impression that it did not work. It *did* work, albeit briefly, but more biochemical attention is needed in order for that treatment to "hold." Simply knowing about a wide variety of treatments can be useful, but those treatments need to be used in context, along with a clear understanding of the biochemical imbalances that could be present, to maximize their potential. So, yes, the rebooting process can be complicated, but it can be done.

Frequency Specific Microcurrent Therapy (FSM)

FSM is one of the most effective neurological rebooting tools, and it's one of my favorites. On my first day of work at Gordon Medical Associates in 2009, I was treated to a lecture by the pioneering chiropractor Carolyn McMakin, who rediscovered and developed this technology. Her lecture caught my attention when she stated that FSM could remove inflammation present in the spinal cord. I was unaware that much of anything could do that. She was discussing a variant of fibromyalgia called *cervical trauma fibromyalgia,* which is caused by persistent inflammation in the spinal cord following a whiplash injury to the spinal cord. She has conducted remarkable studies showing the profound reduction in inflammatory chemicals in the body following a specific two-hour FSM treatment. There was a *twenty-fold decrease* in TNF-alpha, substance P, and interleukin-1 (all markers of inflammation) and a *concurrent twenty-fold increase* in endorphins (our natural painkillers). While the names of these biomarkers may seem obscure, what this means in plain English is that we see a profound, rapid decrease in inflammation and pain in nerve tissue.

I had never encountered anything that worked so quickly and produced those kinds of results. I will admit to a healthy degree of skepticism until I treated my first few patients with cervical trauma fibromyalgia with FSM and saw them get up off the treatment table two hours later with their long-standing pain reduced by 80 percent! And that positive effect lasted. This is a wonderful example of taking a pain signal and literally turning it off—a true rebooting. Dr. McMakin found that while one treatment rarely produced a cure, a series of six or more weekly treatments often resolved this debilitating illness. Just to be clear, by "cure," I mean *complete resolution* of all symptoms of fibromyalgia, which include not only migratory joint pain but also fatigue, insomnia, and cognitive impairment.

Dr. McMakin believes that the precise delivery of the correct frequencies uses energetic resonance to change the structures of key proteins that are "stuck" in a particular pattern secondary to inflammation. She likens it to how a car door opener works: you simply press a button on the key, and it delivers a precise frequency that tells the lock to open, even from a distance. While this seems a bit magical, it simply utilizes the correct frequency resonance to create a physiological effect. A more recent understanding of neuroplasticity (the ability of the brain and nervous system to heal and change) suggests another explanation: the proper electrical stimulation turned off one circuit and stimulated the nervous system to find another circuit to function correctly.

Since becoming familiar with this technology, I have used FSM to treat dozens of patients with cervical trauma fibromyalgia, and most of them have responded dramatically. The process is simple: the patient lies down on a treatment table and is hooked up to a device, which is connected to the patient by electrical wires that are attached to electrically conductive graphite gloves wrapped in warm, wet towels. The towels are placed around the areas to be treated—often the neck and feet. What makes this technology so unique (there are hundreds of electrical devices for pain on the market today) is that it delivers two different frequencies simultaneously. Dr. McMakin has put together a comprehensive list of frequencies that relate to both specific tissues (brain, nerves, spinal cord, ligaments, tendons, muscles, and so on) and physiological impairments (inflammation, congestion, scarring, etc.), which are taught to prospective practitioners in several courses. The correct frequencies are applied while the patient lies comfortably and the practitioner monitors the patient's response. The amount of electrical current is minuscule (a thousandth of an amp), so the procedure is quite comfortable.

As with every treatment referred to in this book, while most patients can handle FSM with ease, exceptionally sensitive patients may need to proceed slowly. A typical session for most patients lasts from thirty to sixty minutes, but sensitive patients may benefit from receiving only thirty seconds to a minute of treatment before being given time to respond and regroup.

Because I am discussing the rebooting of the nervous system here, I am emphasizing the benefits of FSM for these conditions. Any traumatic injury to the nervous system, or nerve pain, may be a candidate for FSM. In addition, FSM can help with detoxification (see Chapter 15). There are specific frequencies for the organs of elimination (for example, the lymphatics, kidneys, liver, skin, and lungs) and for specific toxins (such as from Lyme disease, *Bartonella,* mold, chemicals, or heavy metals) that can enable patients with compromised systems to recover more quickly. I have seen FSM reduce the size of a huge thyroid goiter within twenty minutes of application.

We have just scratched the surface of the potential uses of this modality. For more information and to find a trained practitioner or to learn about using FSM in your own medical practice, go to www.frequencyspecific.com.

The LENS (Low Energy Neurofeedback System)

Another example of a medical breakthrough that is particularly relevant to rebooting the nervous system is the LENS. Biofeedback has been used for many years, in various forms, to help improve body functioning. Most of that biofeedback has involved finding a way to obtain information about a specific biological function (for example, muscle tension, body temperature, blood pressure, or brain waves) and teaching a patient how to alter that information (such as by reducing muscle tension or increasing temperature). It is well documented that this form of treatment can help people relax, lower blood pressure, and relieve migraine headaches and pain.

What makes the LENS unique is that the patient does not have to be aware of the information being processed and does not need to cooperate or work at learning a new skill, which can be very difficult for people with cognitive impairment. To utilize the LENS, a sensor is attached to each ear and another to the scalp. No needles are used. The information detected by the sensors is fed (via an EEG device) into a computer that analyzes brain waves from twenty-one different locations in the brain, a program referred to as *brain mapping*. The resonant, always-changing feedback into the computer through the EEG device is then conveyed back to the scalp, where the brain can detect and recognize the feedback signals. This feedback allows the brain waves to be, in a sense, redirected, creating a literal rebooting of the nervous system.

For example, just recently, a patient of mine who had been very worried about her cognitive decline (she was concerned that she was experiencing Alzheimer's dementia) and had Lyme disease was treated with a second session of the LENS. Within moments of treatment, she was delighted to discover that her mind was much clearer and her thinking had improved dramatically, along with her memory. This shift not only held, but she continued to improve with subsequent treatments. You can see how well this fits into our rebooting model.

The LENS is especially helpful for those individuals who are unable to think clearly because their brains have been damaged or are inflamed by toxins or infections. Most notably, this includes people with traumatic brain injury (TBI), bipolar disorder, chemotherapy or chemically induced cognitive impairment, epilepsy, stroke, anxiety, depression, Tourette's, post-traumatic stress disorder (PTSD), Parkinson's disease, dyskinesias, pseudoseizures, and early-stage Alzheimer's and those on the autism spectrum, which includes ADD, ADHD, and Asperger's syndrome. What all of these conditions share, when uncomplicated by other medical issues, are variants of one fundamental process—*how the brain tries to cope with actual or potential irritability and/or inflammation.* Because of the enormous variety of ways in which people's brains develop, these conditions may present differently in different individuals.

Len Ochs, PhD, who developed the LENS, has studied how the brain attempts to cope with actual or even anticipated irritability for many years. Brain irritability, closely linked to inflammation, can be caused by injury, chemical or heavy metal toxicity, or infections. The brain seems to try to deal with the spread of this irritability by chemically dampening that activity (electrical suppression) and by creating firewalls across both the surface and within the tissues of the brain to prevent those signals of irritation from spreading. *Doesn't this description sound like the neurological equivalent of the cell danger response?*

Though these processes are natural, they interfere with the kinds of communication and connections that the brain needs to function properly. Some people recover from the presence of this altered neurochemistry when the brain puts out the "all clear" signal that the danger has passed. However, others cannot refresh their neurochemistry or reboot the connections in their brains, and they may remain functionally impaired for life.

The LENS, using weak but precisely directed electrical signals to the skin (based on the brain wave analysis obtained from brain mapping), appears to catch the brain's attention to allow it to interrupt its repetitive but nonproductive attempts to block its own communications so that it can resume more normal functioning.

I have found the LENS to be a superb and safe tool for patients whose brain fog and cognitive impairment prevent them from functioning properly. As an example, a young man came to the clinic several years ago after sustaining a brain injury requiring hospitalization for a coma that lasted for several weeks. This injury, which had occurred two years before I saw him, left him exhausted and without the ability to express himself or think clearly, which led to intense depression and mood swings and a fear that he would never recover. Within two months of starting LENS treatments, he reported 60 to 70 percent improvement, after having made little to no progress over the preceding two years despite having undergone a variety of other treatments. His depression and mood swings were gone, his energy levels had improved, and he was able to think much more clearly, enough that he was able to resume college classes.

Of particular relevance for unusually sensitive patients, many of whom have Lyme disease and/or mold toxicity, is the already noted fact that neurological problems may persist even after the toxicity and/or infections have been resolved. These neurological symptoms include dystonias (odd and uncontrollable writhing motions of the arms and legs), pseudoseizures, and tics and spasms. These patients' bodies appear to be stuck in an electrical pattern that they are unable to fix. The LENS has proven a particularly effective tool for many of these patients.

As I do in every section of this book, I must emphasize that very sensitive patients may need to proceed slowly with this treatment. As gentle as the LENS is, short, limited treatments may need to be used initially so as not to overwhelm a highly reactive system. For my sensitive patients, I ask the therapist to start by addressing only "one area at maximal gain," which is code for, "Let's go really easy here."

> " Of particular relevance for unusually sensitive patients, many of whom have Lyme disease and/or mold toxicity, is the fact that neurological problems may persist even after the toxicity and/or infections have been resolved. "

Osteopathic Craniosacral Manipulation as a Model for Treating Structural Imbalances

There is a profound relationship between structure and function. Each influences the other in many ways. An imbalanced or "stuck" structural component in the body will have a significant impact on a person's emotional, energetic, and spiritual life. Conversely, an emotional or spiritual upheaval will change a person's physical structure.

If you slip and twist your knee, for example, you might sustain an injury to any of several structures: your ligaments, muscles, cartilage, and bony elements will demonstrate some degree of damage, and these are often amenable to treatment. However, less appreciated is that when you cannot use your knee as you normally would, after several days you will have to cope with not only pain, but also frustration and disappointment because you cannot participate in the activities you have planned. If your knee does not heal quickly, you may become anxious and/or depressed about whether you have suffered severe damage that might require surgery or leave you incapacitated for a long time.

Similarly, if you come down with bronchitis or pneumonia, you may not appreciate that the accompanying coughing can "jam" your rib heads where they connect with the transverse processes of your thoracic vertebrae. This jamming tightens up your diaphragm, restricting your breathing and thereby causing additional pain and also restricting the flow of oxygen through your lungs, affecting your ability to heal.

To give one more example, if you are anxious about an upcoming IRS audit, you may not realize that your breathing has become shallower and your muscles have tightened up (particularly those muscles that have been injured in the past). You may also lose sleep, which will affect your energy level, and your physical symptoms may augment the anxiety you are experiencing.

Osteopathy is the branch of medicine that specializes in understanding this relationship between structure and function. The founder of osteopathy, Dr. Andrew Taylor Still, recognized it as an organizing principle for diagnosis and treatment and proclaimed, "The rule of the artery is supreme." Metaphorically, what he is alluding to here is that any blockage or restriction needs to be addressed in order for healing to take place. This includes blockages not only in the arteries, but also in the veins, nerves, lymphatic vessels, and any other tissue that is being compressed by a structure that has become disturbed.

This understanding was expanded in the work of Dr. William G. Sutherland, who recognized that the cranium (our skull case) is capable of subtle motion that is influenced by all the tissues inside it, including the supporting ligaments (the dura mater, pia mater, and arachnoid), the cerebrospinal fluid, and the brain itself. This motion of the cranium is directly connected to the sacrum, the triangular bone at the base of the spine, by ligaments that run the length of the spine. Dr. Sutherland taught his colleagues to perceive these subtle movements and redirect them in the service of healing. To emphasize the importance of this component of structure, he referred to it as *craniosacral.*

Sometimes that emphasis on the cranium and sacrum is misunderstood because it seems to focus exclusively on those two structures, but, as the tissues of the body are so interconnected that there really is no separation between them, what we are really talking about is a method of evaluating how the entire body has been influenced by the stressors—physical, mental, emotional, spiritual, and energetic (such as EMF)—to which it has been exposed and with which it must cope. Because the body is a unity of physical, mental, emotional, spiritual, and energetic components, when the healing process gets "stuck," it gets stuck in all of those areas.

The practice of craniosacral treatment can be very sophisticated. On a primitive level, it involves learning how the cranial sutures allow movement of the skull bones (and everything attached to them) and recognizing when normal motion is restricted—and then assisting the body in restoring motion. However, once practitioners can appreciate this process, they learn how to be able to palpate the motion (or lack thereof) of the cranial membranes, the motion of fluids within the cranial structures, and the motion of the brain itself so that they can use that information to direct treatment.

James Jealous, DO, developed what he calls the Biodynamic View of Osteopathy, which includes the perception of even more sophisticated energies and motions. The ability to feel and work with these healing forces allows trained physicians to influence the entire body in profound ways.

I have asked my friend and colleague, Jeffrey Greenfield, DO, to describe the process of what he feels when he provides treatment. As in all fields, there is specific language used that does not readily translate into common usage, but if you can simply follow his description, in his words, I hope you will get an inkling of how deep these perceptions can be and how they can be used in the service of healing or rebooting. Here is Rachel's story, as told by Dr. Greenfield.

RACHEL'S STORY

Rachel was a fortyish-year-old professional in charge of a clinic. Her responsibilities included managing and working with the employees and professionals. She had been treated for Lyme, Bartonella, *adrenal fatigue, and other manifestations of tick-borne disease (TBD). She had mold and mycotoxin exposure but had not yet been evaluated or treated for it. She was quite ill when she was first diagnosed with TBD, but she had recovered greatly from her original condition. Some very skilled physicians had appropriately treated her for some of her conditions, and she thought she felt well. She was still on herbal treatment for* Bartonella *but had come a long way.*

Rachel was familiar with cranial osteopathy and would get occasional treatments from other local osteopaths. She came to me for osteopathic treatment knowing that I had experience treating those with TBD with conventional, alternative, and osteopathic approaches. Her conventional physical exam was quite normal. The evaluation using a biodynamic model of cranial osteopathy was not. Her tissue movement was present, but the force of fluctuation was diminished and a bit disorganized. The fluid fluctuation of her mechanism was also present but was weak and lacked vitality, and it was not connected with her tissue as it should have been. The feel of her potency was scattered. She had a slight medium-frequency vibration, but it was not in the background—which I interpreted to be partially treated TBD.

Toward the end of the treatment, I was working to balance all the effects, pulls, pushes, and movement from the head. As I connected with the feel of the natural world that is around us and in us but is not us, I let her system be part of that. What happened next exemplifies the beauty of osteopathy.

With a small portion of my attention, I noticed the tissue fluctuation, fluid fluctuation, and the feel of the potency go out like a low tide. There was a long pause in all motions. Then, in the center of Rachel's brain, likely in a structure we call the third ventricle, there was a spark, a light, some heat. It felt like a pilot light going on—the health was really beginning to show. Then this spark ignited the fluid, tissue, and potency with a feel of life, health, well-being—this was IGNITION! The tissue, fluid, and potency were all now vivified, in harmony, working together and connected. She took a long, slow, deep breath, and the treatment was over.

In her next visit, Rachel related to me that after that treatment she came out of a fog that she had not realized she'd been in. Her mind was working better, and things were clearer. She noticed that she had been neglecting many of the finer details of her job responsibilities, but she was now able to spot those issues and correct them. She felt awakened, refreshed. She knew she still had Bartonella *and maybe mold toxicity, but she was feeling healthy again. Her system had rebooted itself. Rachel continues to do well and gets treatments regularly, and her treatments hold and her Ignition stays on.*

I hope that Dr. Greenfield and I have been able to give you some idea of the breadth and depth of this form of treatment. The approach provided by craniosacral therapy allows us to enter into treatment very gently, providing an entry into the overwrought nervous and immune systems of our sickest patients and specifically enabling us to directly contact the energetics of mold and Lyme illness to assess its presence and our progress with treatment. Dr. Greenfield has pioneered the use of this modality in helping to diagnose and follow the treatment of patients with Lyme disease and mold toxicity, and he has taught it to dozens of physicians.

This form of osteopathic manipulation has proven to be of great benefit in rebooting the nervous system when forward progress has been stymied. I am honored to have studied with Dr. Jealous so that I could learn to utilize this healing technique for my patients. To learn more, and to find a physician trained in this process, go to www. jamesjealous.com.

Visceral Maniupulation

Another useful technique for rebooting the peripheral nerves was developed by Jean-Pierre Barral, DO, who helped pioneer the technique of visceral manipulation. Courses in visceral manipulation teach practitioners to be able to feel subtle movements of the internal organs (such as the liver, gallbladder, intestines, spleen, and kidneys) and help free up restrictions to motion that have been created by surgical adhesions and injuries. An advanced branch of this training includes the ability to directly influence the peripheral nerves and has proven to be of great value in their treatment. You can learn more and find a list of trained providers at www.barralinstitute.com.

Annie Hopper's Dynamic Neural Retraining System (DNRS)

Annie Hopper, who suffered from severe multiple chemical sensitivities herself, realized that a component of this illness is a limbic system (a component of brain functioning that includes the amygdala) that is "stuck." She developed a non-drug system that might be described as a combination of meditation and structured mental exercises to reprogram, or reboot, the way in which the brain reacts to a wide variety of stimuli.

A major component of the limbic system is to be neurologically alert to anything that could be a threat. When that part of the nervous system becomes overwhelmed by stimuli of any kind—light, touch, sound, smells, foods, or electromagnetic fields (EMF)—it can get stuck on high alert. This is a neurological version of the cell danger response on a cellular level. Once stuck, the limbic system becomes hyperreactive to a wide variety of stimuli.

For many of my most sensitive patients, Hopper's program has been of enormous benefit. I have, as of this writing, utilized it in more than 150 patients and, with just two exceptions, all of them made progress that ranged from good to amazing. When I encounter unusually sensitive patients who cannot tolerate any of the substances that need to be included in their treatment, I strongly encourage them to work with Hopper's program before we try anything new. With few exceptions, after several months on the program, their sensitivity has been reduced to the point that they can begin to move forward.

To learn more, go to www.retrainingthebrain.com. I also encourage patients to read Hopper's book, *Wired for Healing*.

While most of my experience has been with Hopper's program, several of my patients have seen excellent benefits from Ashok Gupta's amygdala retraining program, which is similar. Go to www.guptaprogramme.com to learn more.

I would like to emphasize how valuable this approach is. For patients who have become unusually reactive to almost everything, this retraining is by far the most useful first step I have encountered that will quiet their systems down to the point where we can proceed with more targeted treatments.

Polyvagal Theory: A New Understanding of the Autonomic Nervous System and How to Work with It

What I am going to talk about now is an exciting new understanding of how the autonomic nervous system works. Let's begin by getting a good handle on that name, *polyvagal*. *Poly* is the root word for "many" or "multiple," and *vagal* refers to the vagus nerve, our tenth (X) cranial nerve. New research shows that the vagus nerve (see Figure 9.2), a long and important nerve that goes from the brain to many of our most important organs, including the heart, lungs, and intestines (all the way to the descending colon), is actually made up of two separate nerves that come from two separate areas of the brain. These nerves fuse to become the vagus nerve. We are just beginning to understand that these two parts of the vagus nerve, anatomically called the *dorsal* and *ventral* branches, serve different functions.

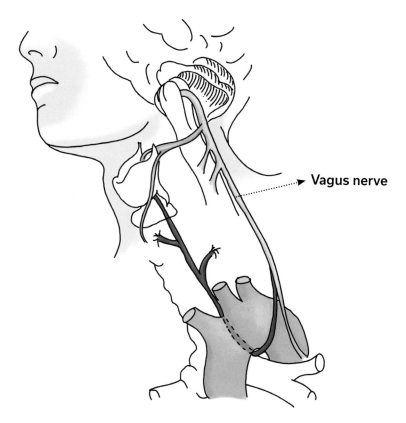

Vagus nerve

Figure 9.2. The vagus nerve is actually made up of two separate nerves, the dorsal and ventral branches, that come from two separate areas of the brain.

> **Many healthcare professionals urge their patients to counterbalance the effects of stress by using breathing and relaxation techniques to stimulate the parasympathetic system.**

Until recently, our understanding was that the autonomic nervous system was divided into two well-defined branches: the sympathetic and parasympathetic systems. You are probably aware that the sympathetic system, activated by stress, is referred to as the "fight or flight" system. The parasympathetic system is thought to balance the effects of stress by relaxing the responses that have been activated by stress. These are primitive responses, following patterns that were etched into the nervous system as human beings evolved.

When threatened by a predator, our pupils and bronchial tubes dilate, our pulse goes up, and our muscles become more reactive—all to enable us to either fight the threat or run away. When we are faced with more modern stressors, like a tax audit, a divorce, or being stuck in traffic and late for work, those same physiological reflexes kick in, but they are much less helpful or adaptive for what we are trying to cope with today.

For decades, this is how we understood this branch of the autonomic nervous system, and often it appears that the sympathetic system is dominant and, at times, out of our control. As modern times became more and more stressful, activation of the sympathetic system contributes significantly to the development of many illnesses. Many healthcare professionals urge their patients to counterbalance the effects of stress by using breathing and relaxation techniques to stimulate the parasympathetic system.

In 1992, after years of studying autonomic responses in his laboratory and realizing that there was another component to the autonomic nervous system that had not yet been recognized, Stephen Porges, PhD, proposed the Polyvagal Theory. Most notable was the discovery that the vagus nerve, a key component of the parasympathetic nervous system, is actually two different nerves coming from two different areas of the brain and fusing to become one. Of particular importance was that one portion of that nerve, the ventral vagus nerve, had evolved to be unusually important in sensing safety or threat in the environment.

In his book *The Pocket Guide to the Polyvagal Theory*, Dr. Porges writes about "a hierarchy in how these neural circuits react to the world. The theory proposes that, similar to the function of the brain in which evolutionarily newer circuits inhibit older ones, the neural circuits that regulate the visceral organs that evolved earlier are inhibited by the evolutionarily newer circuits" (specifically, the ventral branch of the vagus nerve). From an evolutionary perspective, the oldest component of the autonomic nervous system "in mammals is the unmyelinated subdiaphragmatic vagus, which when recruited in defense would shut us down." The next evolutionary stage was "the emergence of a spinal sympathetic nervous system that supported flight/fight behaviors." As mammals continued to evolve, "a new neural circuit evolved that integrated social behavior with the regulation of physiological state...when the new mammalian vagal system is working well, then subdiaphragmatically, the sympathetic and parasympathetic nervous systems are functioning in a homeopathic dance reflecting the positive features of autonomic balance." I will translate this in the following paragraphs to make the profound meaning of these words clearer.

Dr. Porges recognized that humans utilize another stress response when the fight-or-flight response is not working, and that is to "freeze" or become paralyzed. Think of a possum going limp, playing dead until whatever is threatening it decides to go away after detecting no response. Humans do this as well. If you reflect on the reactions you have seen to overwhelming stress, you will probably recognize times when you, or those near to you, have become paralyzed by it. Dr. Porges points out that this response commonly occurs when individuals face physical abuse or wartime attacks, and those who have had to deal with it are often left with a PTSD type of response. He writes that "in response to life threat, both the supradiaphragmatic (ventral branch) vagus and the sympathetic nervous system are down-regulated, enabling the subdiaphragmatic vagus (dorsal branch) to be recruited in defense. The product of this old defense system is immobilization."

The key to understanding the underlying dynamic of this response is that these individuals *no longer feel safe.*

It turns out that the major function of the ventral branch of the vagus nerve is to discern safety, and the importance of safety is the primary focus of the Polyvagal Theory. The perception of safety is hard-wired into our nervous systems. If our nervous systems do not perceive us to be safe, no amount of rationalization will change that physiological response. This is another "systems" way of understanding the cell danger response. Dr. Naviaux's model of how the mitochondria respond to a threat dovetails closely with the Polyvagal Theory. Both models emphasize that the body will continue to operate as if it is under threat until it is convinced that it is safe.

This applies to a wide variety of chronic illnesses. Chronic fatigue syndrome, fibromyalgia, Lyme disease, mold toxicity, and autism are examples of conditions that are difficult to heal because the patient has been under duress for so long that his or her body is essentially saying, "I've been ill for so long that I can't be sure I'm safe, so I won't turn off the distress signal. " And it is hard for the patient to believe that he or she is truly on the road to healing and will once again be safe. Most of these illnesses are characterized by some degree of autonomic dysfunction, meaning that the prolonged strain of a serious illness takes its toll on the autonomic nervous system, and it loses its ability to self-regulate. This can take the form of POTS (postural orthostatic tachycardia syndrome), palpitations, labile hypertension (blood pressure that without warning becomes very high or very low), or anxiety. Therefore, finding a way to help that patient feel safe both physically and psychologically must be an integral part of the treatment process.

It turns out that vagal nerve function is intimately connected to several other cranial nerve functions, especially cranial nerves V (trigeminal), VII (facial), IX (glossopharyngeal), and XI (spinal accessory). Dr. Porges writes, "The myelinated vagus functions as the primary parasympathetic regulator of the organs above the diaphragm. In addition, the area in the brainstem from which the . . . vagus originates is linked to the brainstem area that regulates the muscles of the face and head."

Taken from an evolutionary perspective, Dr. Porges recognizes that these cranial nerves control the movements of the face, neck, and shoulders, among other functions. He has made the fascinating observation that how the upper part of the face (around the eyes and cheeks) moves plays a profound role in how we read the safety and accessibility of those around us. The phrase "his smile did not reach his eyes" is an excellent example: we recognize either intuitively or by careful observation that some people make an effort to smile, but there is no motion of their upper face or eyes, and we have learned not to trust that kind of smile. This is similar to the immobilizing effects of Botox. The expressiveness of our faces, therefore, informs others about what we are really feeling, and they respond accordingly.

The muscles around the eyes are regulated by the VII cranial nerve, and so is a tiny muscle in the ear, called the *stapedius,* which regulates the perception of sound. Dr. Porges writes, "When the middle ear muscles aren't contracting appropriately, we are bombarded by low frequency rumble from background noise. This compromises our ability to understanding human voices." As a result, we tend to withdraw from others.

So, a genuine smile "lights up our eyes," and a forced one does not. Even if we are not consciously aware of what we see in the faces of others, our primitive nervous systems are hardwired to perceive and respond to these expressions.

Therefore, by observing and testing these muscles, we get information about how functional the cranial nerves that control those muscles are. A simple way of doing so is to ask a patient to say "ah, ah, ah, ah, ah" while shining a light into his or her throat. If the movement of the uvula (the dangling tissue at the back of the throat) is minimal or veers off to one side, the vagus nerve is clearly dysfunctional. Another test is to ask the patient to gag, gently. If he or she cannot easily do so, the vagus nerve is likely to be dysfunctional.

Dr. Porges uses this information to emphasize how important it is for physicians to create a treatment room environment that encourages the perception of safety. This includes the color of the walls and the lighting as well as the tone and timbre of the doctor's voice. Patients recognize all of these factors as either safe or not safe and respond accordingly. This is obvious in hospital settings, where constant noises and scanty gowns are antithetical to the perception of safety. Constant examinations and testing (for "vital signs") leaves hospital patients with little sense of dignity or involvement in the decision-making process.

Dr. Porges emphasizes that it is important to reframe the stressful events of a patient's life so the patient recognizes that his or her responses to these stressors have been appropriate and life-saving and must not be thought of as "wrong" in any way. This essential first step in helping the patient redefine safety can be accomplished by emphasizing that however the patient coped with intense traumatic events in the past, those efforts succeeded, because he or she is now sitting, alive, in the exam room. A perception of safety begins with a reminder of how valuable those coping efforts have been rather than a focus on how stressed the patient has become. With an appreciation of the adaptive value of how he or she has coped, the patient can consider the possibility that learning new coping strategies may allow a move forward into healing.

Understanding the importance of the perception of safety and its role in healing the autonomic nervous system allows us to approach treatment with more precise and specific efforts. What I have provided here is a simplified version of this important theory; I encourage you to read Dr. Porges' book.

I would also like to recommend a book by Stanley Rosenberg called *Accessing the Healing Power of the Vagus Nerve.* It provides an excellent overview of Dr. Porges' theory and gives a series of simple exercises that quiet the ventral branch of the vagus nerve and the associated other cranial nerves. These exercises take only five to ten minutes to do, and my patients are already reporting benefits

> **Understanding the importance of the perception of safety and its role in healing the autonomic nervous system allows us to approach treatment with more precise and specific efforts.**

from doing them daily. Rosenberg utilizes craniosacral manipulation (see page 152) to accompany his exercises, which is a wonderful combination.

FSM (frequency specific microcurrent) can help quiet the vagus nerve as well. It uses gentle electrical stimulation and has proven very helpful in treatment (see page 147).

When we contemplate rebooting the nervous system, taking the autonomic nervous system into account is important. I have found that for my most sensitive patients, combining Annie Hopper's DNRS program with an understanding of Polyvagal Theory has been of particular value in getting treatment back online.

Acupuncture

Whole volumes have been written about acupuncture and its potential benefits. Here, I want to emphasize that there are several ways in which acupuncture can help reboot a nervous system that is "stuck."

It is well documented that one of the effects of acupuncture is to stimulate the body's production of endorphins. When a patient has been wrestling with pain for a long time, rebooting the ability to produce endorphins can be of great benefit.

While the traditional medical curriculum does not embrace the subject of energy, it is increasingly clear that a wide array of natural energies, with different frequencies, are an integral part of normal human physiology. When patients' physiologies are seriously compromised, their energies and emotions appear to be compromised as well. The practice of holistic medicine has long emphasized the integration of mind, body, and spirit, and it is in patients with chronic illness that we see the need for this integration most clearly. When the physical body is stuck, energy systems and emotional and spiritual systems almost certainly will be stuck as well.

One of the basic principles of acupuncture is teaching practitioners to feel and mobilize energy in the body. Many of my patients have responded well to acupuncture when nothing else was helping. However, I have also found that the response to acupuncture does not last as long as we would hope. The techniques referred to earlier, such as FSM and LENS, appear to reboot the body at a deeper level, and their effects tend to "hold" better.

For the following somewhat more detailed understanding of how an acupuncturist is trained to understand this process, I am indebted to Lauren Gordon Fahn, MAOM, LAc. Her description of how pain persists in the body because of an inability to resume its normal flow pattern adds another dimension to our discussion of "stuck" systems.

• • • •

In Traditional Chinese Medicine, acupuncture is known to be an adjunctive therapy. In Western medicine, acupuncture is often seen as a treatment on its own. Acupuncture alone is very effective in treating acute pain in a moderately resilient body. Chronic pain and chronic illness are often refractory when treated with acupuncture as a stand-alone treatment. The following will explore why acupuncture as a stand-alone treatment has limitations in treating chronic pain and chronic illness and how drawing on some of acupuncture's more foundational tenets can improve acupuncture's efficacy as an adjunctive treatment for chronic issues.

Acupuncture opens and moves energy that is already in the body. It does not add energy to the system directly, but supports organ systems, known as yin organs, *that are capable of adding to the body's energy by bringing in food, air, and water and transforming these materials into energy.*

The nervous system is often delayed in repair after chronic pain and illness. This can be due to problems of depletion and to residual energetic blockage causing chronic inflammation, which inhibits nourishment and repair.

There are two major types of energy in Chinese Medicine: deficiency type and excess type. Deficiency type is from depletion of energy sources, while excess type is from inflammation.

It follows that there are two basic kinds of pain in acupuncture: pain that is caused by deficiency and pain that is caused by excess. This is a very simple perspective, as there are many other attributes of pain in Chinese Medicine.

In the simplest terms, deficient forms of pain feel hollow, dull, vague, and often constant and feel better with applied pressure. This kind of pain occurs when an area of the body does not have enough energy flowing to it or has been overused for some time and the channel itself does not have enough energy.

Excess forms of pain feel sharp, extreme, pulsing, and often very tight; this kind of pain is usually worsened by pressure. Chinese Medicine sees most excess pain coming from an exterior source, an energy that the body itself couldn't benefit from that has con-gested an area.

Most pain starts out as excess pain from a major sudden event. Over time, the dynamic between excess and deficiency pain arises as the body becomes worn down and has not been able to rid itself of the excess energy that is an intrusion on the system's healthy energy flow. Therefore, chronic pain is often an excess and deficiency condition.

When the body does not have enough energetic resources, ener-gy cannot flow smoothly. Smooth-flowing energy is what keeps the body healthy and pain-free. Inadequate energy and therefore inadequate flow of energy leaves voids in the body that are easily invaded by exterior excess energies. Often the body does not have energetic resources to move out this excess invasion. These block-ing invasions are felt as excess-type pain.

Excess-type pain can be treated by acupuncture very easily in a resilient body. If the invading energy resides for too long and depletes the body, or if the invading excess energy invades an already depleted body, the deficiency that created space for the excess energy to invade remains an issue. Chronic pain patients therefore usually experience deficiency forms of pain as well as excess forms of pain. Chronic pain can become a vicious cycle of depletion and invasion, of excess and deficient pain.

When a patient is somewhat resilient, acupuncture treats the deficiency issues by treating the area that needs support. This is functional when the source of the issue is still a source of energy in the body. In modern society, chronic illness lasts longer because we have better resources available to us. When you try to directly treat an issue area that is overly depleted, you are "tapping an empty well." This means that using a different approach when treating chronic pain and a depleted nervous system is important.

From an acupuncturist's perspective, a depleted nervous system is prone to blockages from excess-type energies. These blockages are often seen as inflammation in Western medicine. Clearing the cycle of depletion and blockage is key to rebooting the nervous system. We will focus on the concept of deficiency and excess pain because it is easier to explain the treatment principles that are im-portant to restarting the system and breaking the depletion cycle. But, pain or not, rebooting the nervous system after chronic health issues benefits from returning to four basic tenets of acupuncture treatment to assure greater efficacy of acupuncture in treating chronically ill patients.

The most basic principle of acupuncture is that acupuncture is not a primary or "high" form of treatment, but only an adjunctive treatment. "High" forms of medicine are herbs, food, breathing exercises, and different forms of physical movement. In Traditional Chinese Medicine, these are all primary ways of adding energy directly to the body. The approach to treating chronic pain with acupuncture must focus on assuring that the excess issue is expelled and the deficiency is built up. This rebuilding takes place not by needling the depleted energy channels but by employing other forms of treatment that are capable of rebuilding the body's energy.

The classic Chinese Medicine tenet of treating the part of the body that is healthy should be a focus in approaching the chronically inflamed. Chinese Medicine sees the body in many different forms of duality. The most integral form is yin and yang and the pairing of yin and yang energy/organ channels. The Chinese see the body as different types of energy that are categorized by organ strengths and are named after organs. Yin organs are organs that help the body create energy out of water, air, and food, while yang channels move the energy produced of their yin partners. Most chronic issues are issues of the yin organs.

The principle of treating the stronger or healthier half when treating yin illness is to treat the yang partner channel of the impaired yin area. A simplistic definition of yin illnesses is where the body is depleted, fatigued, heavy, foggy, and sluggish. Yin organs include the spleen, lungs, kidneys, and liver. In Chinese Medicine, yin illnesses are often digestive, issues with general vitality, neurological issues, and chronic immunity issues.

The final tenet to come back to when treating chronic pain is to move energy blockages before treating deficiency within the span of a treatment session. This can mean treating the excess conditions first and opening up yang channels that enable energy movement in the body that has been stagnant in the yang channel itself or in its yin partner.

In summary, when treating chronic pain patients, there are usually some very basic tenets to keep in mind. Chronic pain patients have blockages that inhibit the natural flow of energy/electricity and nutrients. Low levels of nutrients degrade the body's neural infrastructure. When neural pathways are not maintained, they do not allow the proper generation of electrical messages. Acupuncture's functionality is dependent on having energy to move and passageways that are capable of moving energy. This means that acupuncture is best used alongside other treatments that have the ability to build up the body at the same time the acupuncture is expelling excess pain and opening up areas that are energetically blocked.

● ● ● ●

Interference Fields

Interference fields are a newly recognized impediment to healing. These electrical fields are generated by old scars, injuries, and the residua of dental procedures that leave certain bodily energies "blocked" and unable to heal. Following the discussion of acupuncture, I hope you can appreciate these energetic disturbances. Once they are identified, methods can be applied to release these blockages and promote healing. You will quickly see how the understanding of interference fields creates the potential for rebooting. I am appreciative of my friend and colleague, Dave Ou, MD, from Atlanta, Georgia, who is more expert in this area than I am, for providing the following discussion.

● ● ● ●

One little-known concept in integrative medicine is the treatment of interference fields. It comes from a modality developed in Germany called neural therapy. *It was developed by two German brothers, Walter and Ferdinand Huneke, MD, almost 100 years ago. Dr. F. Huneke was treating a woman with a frozen shoulder with injections of procaine, an anesthetic similar to the lidocaine used by dentists today. The weekly injection usually helped the woman's shoulder pain, but one day it did not. She mentioned that a scar on her left shin was bothering her, so Dr. Huneke decided to inject the scar with procaine. To the amazement of Dr. Huneke and the patient, the shoulder pain vanished instantly.*

Dr. Huneke wondered why injecting one part of the body would affect another part. These areas would eventually be called interference fields *because they appeared to interfere with the normal function of other parts of the body. It is hypothesized that this phenomenon is regulated by the autonomic nervous system, which is why the modality of injecting procaine is called* neural therapy.

Researchers found that interference fields were characterized by an abnormal resting cell membrane potential. In most healthy cells, there is a −70 mV difference inside the cell relative to outside the cell. As a cell becomes sick, the resting cell membrane potential starts to change. The further it deviates from −70 mV, the less oxygen and nutrients enter the cell and the fewer waste products are excreted.

If you ask a pharmacologist how anesthetics like procaine and lidocaine work, he or she would tell you that they temporarily disrupt the resting cell membrane potential of the nerve cells that transmit pain signals. When nerve cells lose their cell membrane potential, they cannot transmit pain signals to the brain. When the anesthetic wears off, the resting cell membrane potential returns to normal, and it becomes able to transmit pain signals again. It was assumed that if a cell had an abnormal resting cell membrane potential before an anesthetic was given, then the membrane potential would remain abnormal after the anesthetic wore off. However, researchers found that the cell membrane potential of those abnormal cells would improve by returning closer to −70 mV after the anesthetic was given.

Drs. Huneke spent the rest of their careers figuring out where procaine could be injected to create therapeutic effects. The injection of procaine into interference fields is not the only way to treat them. In his textbook Neural Therapy: Applied Neurophysiology and Other Topics, *Robert F. Kidd, MD, writes that a device called CAMS (Crosby Advanced Medical Systems) has replaced his use of procaine. He reports that it is as effective as procaine injections but is noninvasive and painless. I have found the same in my own practice. Note that there is little research on how CAMS works, and it works better in the hands of those who are experienced with energetic medicine modalities such as Qigong or Reiki or who are able to perform bioresonance testing, such as autonomic response testing (ART). Dr. Kidd writes that certain types of osteopathic manipulation and energy psychology are also effective. I have also found that frequency specific microcurrent therapy (FSM) can successfully treat many interference fields, and I am currently exploring whether biomagnetism or biomagnetic pair therapy by Dr. Isaac Goiz Duran can do the same.*

The most common interference fields addressed by neural therapy (or noninvasive neural therapy, such as by using CAMS or FSM) can be classified into three types: scars, ganglia, and organs. Scars can be the result of surgical procedures such as skin biopsies, appendectomies, cholecystectomies, and tonsillectomies. They can also result from cuts, piercings, and tattoos. However, not every scar becomes an interference field. Only the ones that develop an abnormal resting cell membrane potential become interference fields. In the language of Traditional Chinese Medicine, these scars are thought to block the flow of Qi, or energy, through meridians. Dietrich Klinghardt, MD, PhD, says that scars and other interference fields can cause a phenomenon called switching. *With switching, treatments that usually work well can cause unexpected negative effects or have no effect at all. In my experience, people with scar interference fields tend not to respond well to most treatments, including vitamins, herbs, and acupuncture. Once the scars are successfully treated, the patient will respond better to these other treatments.*

The second major category of interference fields addressed with neural therapy are autonomic ganglia. Autonomic nerves are the nerves that control tissues and organs such as blood vessels, the brain, heart, lungs, and digestive organs. As they leave the brain, the autonomic nerves form clusters called ganglia *before branching out to tissues and organs. For example, the superior cervical sympathetic ganglia located in the neck control everything in the head and the blood vessels to everything in the head, including the brain, hypothalamus, pituitary, and pineal glands. If the superior cervical sympathetic ganglia become an interference field, the brain, hypothalamus, pituitary gland, and pineal glands and the blood vessels feeding them will not work properly.*

One common scenario is a person with fatigue that is thought to be from adrenal fatigue, but who does not respond to adrenal adaptogens, such as ashwagandha or ginseng. Since the pituitary gland regulates the adrenal glands, the problem could be from the pituitary gland. The problem with the pituitary gland could be due to blood flow to the pituitary, which is controlled by the superior cervical sympathetic ganglia. Correcting this interference field would be the key to improving adrenal function in this case.

Likewise, because the superior cervical sympathetic ganglia controls blood flow to the brain, an interference field in this ganglion can cause a whole host of brain symptoms, such as fatigue, headaches, mood problems, and poor memory and concentration. Treating this interference field will reverse those symptoms. Other frequently treated autonomic ganglia in neural therapy include the retrobulbar, sphenopalantine, submandibular, otic, thoracic sympathetic, celiac, lumbar sympathetic, and inferior hypogastric ganglia.

The last major category of interference fields is tissues and organs. Within this category, the most commonly treated areas are the teeth, tonsils, and sinuses. Dr. Klinghardt has called the tonsils "the toilet of the brain." When waste products are excreted from the brain, they enter a system called the glymphatic system. This system is a lymphatic pathway that includes other lymphatic tissues, such as the tonsils. If the tonsils become an interference field, waste products can back up into the brain, causing brain symptoms such as fatigue, headaches, poor concentration, and mood disorders. For many people with brain symptoms, investigating and treating any tonsillar and cervical sympathetic ganglia interference fields is important.

In the chronically ill, interference fields in the teeth or where teeth used to be are common. It is believed that each tooth sits on an acupuncture meridian. For example, the wisdom teeth connect to the heart and small intestine meridians. Interference fields that develop after wisdom teeth are removed can contribute to heart and digestive disorders. When someone with digestive symptoms is not responsive to common treatments such as probiotics, enzymes, and avoidance of food allergens, a dental interference field should be considered.

There are three major causes of dental interference fields: mercury amalgams, cavitations, and root canals. Mercury amalgams, often called "silver amalgams," are composed of over 50 percent mercury, a known toxic heavy metal. These fillings appear gray or silver as opposed to "white." Volumes about the health hazards of mercury amalgams have been written, but this discussion will be limited to their role in creating dental interference fields. If a mercury amalgam becomes a dental interference field, neural therapy can help temporarily but will not resolve it. The amalgam would have to be removed, ideally by a biological dentist who is trained and uses protocols for the safe removal of mercury. Be careful of dentists who claim to be biological dentists or to have training but do not use the proper protocols.

The very sick should not undergo amalgam removal unless they are under the care of a highly qualified healthcare professional. I typically like to treat my patients for at least a few months to stabilize their health before recommending that they undergo amalgam removal. The reason is that no matter how careful a good biological dentist is, some mercury can leak during the process, and that can trigger high levels of inflammation. I also like to closely monitor patients during and after amalgam removal. Over the years, I *have seen chronically ill patients who say that their illness began soon after amalgam removal. At the same time, I've seen many patients have major health breakthroughs after amalgam removal. The keys to optimal results are timing, preparation, and close follow-up and management.*

Cavitation is a term that many people, including dentists, are unfamiliar with. It should not be confused with cavities, which are breakdowns of the enamel. A cavitation is an area of dead bone in the jaw, also known as osteonecrosis. One of the most common locations in which this problem can develop is the site of a tooth extraction; if the bone did not get enough blood flow to heal completely, dead bone can be left behind. Wisdom tooth extraction sites are the most common sites that can become cavitations. Cavitations also can form where root canals have been performed.

Only a subset of biological dentists are knowledgeable about addressing cavitations. The best current technology to detect them is a 3D Cone Beam Scan when read by a biological dentist who is trained to look for them. They can also be found by someone who is highly skilled in the use of autonomic response testing, as developed by Dr. Klinghardt, or EAV (electroacupuncture according to Voll). A practitioner or dentist with mediocre skills may not get accurate results.

For treating cavitations, one of the most common options is to inject medical ozone into the gums over the cavitations. This treatment helps most, but not all, of the time. The other common option is surgical curettage, which is a surgical procedure similar to draining an abscess in the jawbone. It has been reported that using a LaserCAMS, from Crosby Advanced Medical Systems, can help clear interference fields caused by cavitations. I was skeptical that such a device would be successful, but my initial observations have shown it to be successful in my own practice, even in those who failed ozone therapy.

The third major cause of dental interference fields is root canals. In his book Root Canal Cover-Up, George Meinig, DDS, wrote that it is impossible to completely sterilize a root-canaled tooth, so bacteria always remain. If the bacteria grow to a certain extent, the tooth can become an interference field. The only definitive treatment is extraction of the tooth. Ozone or the use of a LaserCAMS can sometimes help temporarily.

A fifty-five-year-old woman came to my practice for vulvodynia and interstitial cystitis, painful conditions of the vulva and bladder for which science does not know the cause. For this patient, they had started about eight years ago, along with pain in her right leg and trouble speaking and thinking. She had been seen by numerous urologists and gynecologists. She was eventually diagnosed with chronic Lyme disease and treated with ten months of IV antibiotics. They helped her brain symptoms but not her bladder, pelvic, and leg pain. She consulted me after receiving the antibiotics. I found four cavitations where her wisdom teeth had been extracted decades earlier. Treating them with a LaserCAMS resulted in the elimination of her leg pain and an 80 to 90 percent decrease in her bladder and pelvic pain. I also found persistent Lyme and co-infections. Over the years, I have seen many patients who were not able to fully recover from chronic Lyme until their dental interference fields were resolved.

As another example of the importance of interference fields, a woman in her fifties came to me several years ago with fatigue, weakness, multiple chemical sensitivities, headaches, brain fog, and sensitivities to most supplements and medications. We discovered that her home was moldy, and it took a few years before she could move into a non-moldy home with modest improvements. We then realized she was sensitive to electromagnetic radiation, but it took a couple of years before she and her family believed it enough to remove wireless technology, which resulted in additional modest improvements. She had multiple amalgams, root canals, and cavitations in her jaw, but she was too weak to even visit a dentist; even riding in a car for a few minutes left her exhausted. Trying to find a way to get her stronger was challenging until we introduced CAMS to clear her interference fields. We used it to address numerous scars on her body and her cervical sympathetic ganglia. She reported the most improvement when we used it to treat the interference fields caused by her cavitations, root canals, and amalgams. Slowly, she became stronger and less chemically sensitive such that she was able to tolerate longer and longer car rides, including to my office, and to go outside to do some gardening. For the first time, she was able to visit friends and family who lived a few hours away without becoming exhausted. We hope that she will soon be strong enough to see a biological dentist to remove her amalgams to take her health to higher levels.

• • • •

For more on neural therapy, please see the Suggested Reading section at the back of this book. Neural therapy classes can be found at www.klinghardtacademy.com.

Sensory Stimulation: Sound

Depending on the specific causes of a "stuck" nervous system, any stimulus, if administered properly, theoretically is capable of rebooting that system. In the Zen tradition, for example, there are many stories of a Roshi (teacher) working with a student who suddenly slaps a sandal on the floor, and this unexpected stimulus jolts the student into an awareness that he or she was previously unable to attain. In many traditions, chanting and singing are used to facilitate healing. Individuals have reported an unexpected connection to the "music of the spheres" that has profoundly changed their perception of the world.

My friend and colleague Joel Friedman, MD (who describes the use of ketamine in Chapter 19), has studied the Australian instrument the digeridoo and uses the vibrations he elicits from it to create healing shifts in the nervous system for many of his patients. See also the reference on page 178 to the Electronic Ear and other treatments that use music and vocalization to reboot the nervous system.

The ReCODE (Reversing Cognitive Decline) Program

A method for rebooting a chronic illness that has long been thought of as unresponsive to treatment can be found in the recently published and groundbreaking book *The End of Alzheimer's,* by Dale Bredesen, MD. Dr. Bredesen reviews the research that has given us a new understanding of Alzheimer's disease and new hope for successfully treating it. It has long been known that a buildup of amyloid plaque is present in Alzheimer's dementia, but the medical profession has placed its focus on treating the plaque rather than the causes of its accumulation. Dr. Bredesen describes three types of Alzheimer's; a given patient may have one or more of these components to his or her illness:

1. Inflammation (from infection, diet, or other causes)
2. A shortage of nutrients, hormones, and other brain-supporting substances
3. Toxic substances, most commonly mold toxins and heavy metals

Dr. Bredesen has published research that demonstrates cures for Alzheimer's disease by addressing *all* of these factors, and well over 200 of his patients have responded well to his program. It may not come as a surprise to you that many of the components of that program are discussed in detail in this book and are part of our understanding of chronic illness. His application of this information to Alzheimer's specifically is welcomed and clear.

Adjunct Therapies That Can Improve Rebooting Efforts

While FSM, the LENS, and biodynamic osteopathy are the rebooting therapies that have changed the lives of my patients the most, there are many other therapies that can be used to improve neurological functioning so that rebooting efforts can be more effective. These include both electrical and biochemical approaches, and I want to mention some of the most common (and often overlooked) ones here.

Magnesium Supplementation

Magnesium deficiency is common to most chronic illnesses. This mineral plays a key role in all nerve and muscle functions and is an essential cofactor in hundreds of biochemical reactions in the body, many of which relate to how we generate energy.

Symptoms of magnesium deficiency include fatigue, weakness, muscle cramping and spasms, pain, cognitive impairment, cardiac arrhythmias, insomnia, and depression. A particularly clear symptom of magnesium deficiency is that manual treatments, such as massage, chiropractic, or osteopathic manipulation, "do not hold." This means that there is initial improvement, but within a few hours, all the benefits have worn off.

Accurately measuring magnesium levels can be more difficult than you might think. Magnesium is found primarily within the cells (intracellular), so blood tests for magnesium may not detect a significant deficiency because they measure what is outside the cells (extracellular). Despite this fact, most physicians measure magnesium with a simple blood test, and many patients are informed that their levels are normal. To get an accurate reading, you need to measure intracellular magnesium levels, which can be obtained by testing red blood cells (which can be done, if ordered, by most laboratories) or, even more accurate, skin cells. The latter can be achieved by scraping some cells from the tongue (which my patients refer to as a non-painful "Pap smear of the tongue") and smearing them onto a slide that is sent to a lab that can analyze the contents for magnesium. I have used Intracellular Diagnostics for this measurement for twenty years, and it is the most accurate assessment I know of.

Based on the results of testing, deficiencies are usually treated with oral magnesium products. The form of the magnesium matters a great deal, because the most common supplements, such as magnesium oxide, are very poorly absorbed. I have used magnesium taurate and magnesium glycinate with great benefit over the years. If the deficiency is profound, with intracellular magnesium levels measuring below 31.0 mEq/L, oral magnesium may not be sufficient, and intravenous magnesium administration may be necessary.

Magnesium can be absorbed to a lesser extent directly through the skin. Using Epsom salt baths can help increase magnesium levels and is also helpful for detoxification and relaxation. Magnesium lotions also have been helpful in improving magnesium levels and relaxing muscles.

The main point I am hoping to impart here is that without adequate levels of magnesium in the body, rebooting efforts may not be effective. We must look at everything we attempt as part of a bigger whole.

Intravenous Phosphatidylcholine

Phosphatidylcholine was reviewed in the discussion of treatment for mold toxicity (see page 81), but I want to mention it here, too, because it belongs in this discussion as well. Phosphatidylcholine is a major component of all cell membranes and of the myelin sheath (the protective covering of a nerve cell). Giving phosphatidylcholine intravenously has been a wonderful adjunct treatment, helping to strengthen and heal nerve tissues and membranes. Again, sensitive patients may need to start with minuscule doses (0.5 to 1 cc) and work up slowly.

Cranial Electrical Stimulation (CES)

CES is form of TENS (transdermal electrical nerve stimulation), a technique commonly prescribed for the treatment of chronic pain, but it operates at different frequencies, using a different conceptualization so that it can be used on the head, unlike traditional TENS units. Daily twenty-minute treatments have been shown to markedly improve neurotransmitter levels and relieve depression and pain. This is not, technically, a rebooting, but rather a use of electrical stimulation to enhance the body's ability to make the proper neurotransmitters. However, I have had many patients respond to this non-drug approach with excellent results over the years.

Amino Acid Supplementation

In multiple research studies, Marty Hinz, MD, has shown that the use of 5-HTP (5-hydroxytryptophan, the precursor to serotonin) and L-tyrosine (the precursor to dopamine, epinephrine, and norepinephrine), when used with cofactors and other amino acids, especially cysteine, dramatically improved a wide variety of illnesses. This is particularly true of neurodegenerative diseases, such as Parkinson's, but applies to depression, anxiety, chronic fatigue, and fibromyalgia as well. Dr. Hinz has pioneered a method of measuring the levels of these neurotransmitters in the urine and then using the results to provide supplements of amino acid combinations for treatment that produces excellent results.

Addressing Kryptopyrrole Buildup

This somewhat intimidating term simply refers to a group of chemicals that are involved in the process of breaking down red blood cells once those cells have outlived their purpose. You may recall that red blood cells live for only about ninety days, after which time they are recycled. Sometimes there is a glitch in the way hemoglobin, an essential component of every red blood cell, is broken down, and we get an accumulation of kryptopyrroles, which can lead to anxiety, depression, difficulty dealing with stress, and other symptoms. This glitch can be genetic, but it is frequently triggered by mold toxicity and Lyme disease, so we see it much more commonly in patients who are unusually sensitive.

Pyroluria (another name for it) can easily be measured with a simple urine test. The good news is that it basically reflects a deficiency of zinc, vitamin B6, and magnesium and is relatively easy to treat by supplementing with those nutrients. This biochemical imbalance is surprisingly common in chronically ill patients, and missing this diagnosis (by neglecting to test for and treat it) also denies us the opportunity to provide a simple, benign treatment that can help quiet down patients' already overstimulated nervous systems.

Correcting Zinc and Copper Imbalances

There is a reciprocal relationship between zinc and copper in the body, and many patients who present with psychological symptoms have an excess of copper and a deficiency of zinc. Any laboratory can measure plasma zinc and serum copper. If an imbalance is found, it is relatively easy to correct by increasing zinc intake and eliminating sources of copper exposure (such as drinking water or supplements) that are creating the imbalance.

MCT Oil and Coconut Oil

A combination of medium-chain triglyceride (MCT) oil and coconut oil can help reduce brain inflammation in many patients. This approach was pioneered by Mary Newport, MD, in the treatment of her husband's Alzheimer's disease. My treatment protocol, below, is easy to do, tastes pretty good, and has few side effects (mainly diarrhea if too much is taken). This protocol can be particularly helpful for patients with hypoglycemia.

1. Mix 16 ounces of MCT oil with 12 ounces of organic coconut oil. You will need to liquefy the coconut oil first by placing the bottle in hot water or placing the coconut oil in a double boiler until it can be more easily mixed with the MCT oil. Shake well to combine the oils thoroughly; the mixture will become a thick slurry.

2. Start by taking ½ teaspoon of this mixture one to three times a day and *slowly* increase to 2 tablespoons three times a day. (It works much better if you can take it three times a day or even more often.) Shake well before each use.

3. Be sure to take the last dose in the evening.

4. Store the mixture at room temperature to keep it in a more liquid/usable form. *Do not put it in the refrigerator.*

Diagnosing and Treating Hypoglycemia

Hypoglycemia simply means low blood sugar (*hypo* = low and *gly-cemia* = sugar in the blood). Low blood sugar is quite common in my chronically ill patients. Its presenting symptoms are fatigue, brain fog, palpitations, and anxiety.

For reasons I do not understand, the medical profession turned its back on the concept of hypoglycemia in the 1980s and has not embraced its reality. Accordingly, very few physicians administer the requisite four- or five-hour glucose tolerance test needed to diagnose hypoglycemia, and many laboratories have stopped doing it. Despite this, hypoglycemia is very real and is easily diagnosed. More importantly, treating it can improve the underlying physiology. Key components of treatment are following a high-protein, low-carb diet; snacking on high-protein, low-carb foods; taking 200 milligrams of chromium picolinate twice daily; and using the MCT/coconut oil protocol outlined on the opposite page. (The ketogenic diet has been of particular benefit here.) I will reiterate that missing this diagnosis also means a missed opportunity to provide a simple treatment that can help a patient heal, and feel better, more rapidly.

Addressing Methylation Dysfunction

Methylation is an important component of the cell danger response, and the vast majority of my patients have difficulty with methylation. I discuss it in more detail in Chapter 14, but I want to emphasize here that unusually sensitive patients may not be able to address methylation until they have worked on removing toxins from their bodies.

Neuroplasticity Research

Annie Hopper notes that some of the research on which her DNRS program (see page 156) is based is found in the books *The Brain That Changes Itself* and *The Brain's Way of Healing,* both by Norman Doidge, MD. Reading these books has been inspiring to me, as they emphasize how much we are learning about the brain's ability to heal, once injured, from a wide variety of stimuli, including visualization, sound, light, and electrical stimulation. While I do not yet have personal experience with the following modalities, I would like to mention some of the most promising ones so that you can explore these possibilities in more detail if you wish:

- Michael Moskowitz, MD, a pain specialist from northern California, has pioneered specific visualizations that have led to the relief of chronic pain even when that pain has been present for decades.

- Fred Kahn, MD, a Toronto vascular surgeon, became impressed with the healing power of specific frequencies of light. He uses a process called BioFlex Laser Therapy to help with wounds, pain, and cognitive dysfunction.

- A research team at the University of Wisconsin–Madison developed the PoNS device, which is a thin disc placed on the tongue that delivers specific gentle electrical frequencies, combined with exercises designed specifically for each patient, to stimulate healing of the brain and nerves. The research has been completed, and a new company, Helius Medical Technologies, plans to make this device available commercially within the next year.

- The Electronic Ear, a device developed by Alfred Tomatis, MD, and further evolved by Paul Madaule, uses filtered music and vocalization to change the way sound is perceived. The LiFT Listening Fitness Trainer and a similar device, the Integrated Listening Systems (iLs), have been shown to heal children with autism, auditory processing disorders, ADD, and ADHD and are showing promise in the treatment of many other neurological dysfunctions.

- Michael Merzenich, PhD, has developed computer programs to preserve the plasticity of the brain; the software is available through Posit Science.

- Constraint-induced movement therapy, developed by Edward Taub, PhD, has enabled patients who have had strokes, even long-standing ones, to improve and heal neurological function by encouraging the use of the nonfunctional limb.

- In his book *Brain Lock,* Jeffrey Schwartz, MD, demonstrates how consistent visualizations can heal obsessive-compulsive disorder (OCD) by using a relabeling and refocusing process.

As more and more scientists get excited about our evolving understanding of neuroplasticity, I anticipate an explosion of new ideas, techniques, and technologies that will help cure patients who were previously thought of as permanently damaged.

Apart from these, please keep in mind that I have included only the rebooting methods with which I have had direct experience. Many others exist, and many more are in development. It is my intention to bring this large body of information to your attention so that it will add to the message of hope that no matter how long you have been ill, there are multiple healing modalities available to help you.

Even if you have tried one or two of these therapies without success, don't give up! There is no way to know in advance which technique will help which patient. Finding the right solution requires a trial-and-error process and a willingness to keep searching for answers.

CHAPTER 10

Rebooting the Immune System

Every Day Is Groundhog Day

I don't think the immune system is fully appreciated for the efforts it continuously expends to keep us well. I often hear cancer patients comment, "I never even get a cold. How could I have gotten cancer?" The immune system is not a simple or singular entity, but it includes the parts that *separately* deal with allergens, find and eliminate cancer cells, and find and eliminate infectious microbes. While there is some overlap between these functions, to a certain extent they are not connected. That is why you can get cancer even if you have never had a cold—those are different immune functions. When a supplement is promoted as being "good for your immune system," that phrase leaves me baffled; which part, exactly, is it good for? These generalities do not help the consumer. If you have a viral infection, for example, you should know whether a particular supplement is good for that specific infection. (In fairness to the supplement industry, the FDA does not always permit companies to be so specific in their product labeling unless very strict research standards have been met.)

The immune system has to play both detective and gatekeeper. It is constantly scrutinizing everything that enters the body for possible toxicity or infectiousness, or, to be more global, "foreignness"—meaning that it must be ever alert that nothing "not us" gets past its defenses. If, somehow, something does threaten us, the immune system then must mobilize its full resources (white blood cells of various kinds; the complement system, which is a part of what is called the *innate immune system;* detoxification systems; and antibodies) to eliminate the threat as quickly and efficiently as possible. This is a

Herculean task, and one that is rarely accomplished perfectly. With the massive increase in the number of chemicals to which we are now exposed (many of which did not exist fifty years ago), it should come as no surprise that the detailed structures of some of these chemicals are actually quite similar to the chemical structures of substances found inside our bodies. Our immune mechanisms are constantly making many decisions about what to fight and what to allow inside. As you might imagine, this opens the door to mistakes. Given the massive amount of work our immune systems are forced to do, it is amazing that we manage to function at all.

A central concept here is that the immune system uses an inflammatory response to do its job. When you get a cold or viral infection (or fight cancer), your immune system recognizes the threat and brings to bear a wide variety of white blood cells (such as neutrophils, monocytes, lymphocytes, and eosinophils) and antibodies to deal with it. These white blood cells produce cytokines, which are messenger molecules that set into motion a cascading inflammatory response that is intended to destroy the invader.

Many patients are under the false impression that when they feel "flu-ish," with chills, fever, muscle aches, and brain fog, they are feeling the virus at work. Actually, they are feeling the immune system *fighting* the virus and are experiencing that flood of inflammatory cytokines that are designed to do the job.

We are increasingly aware that the central feature of many chronic illnesses, including fibromyalgia, myalgic encephalopathy/chronic fatigue syndrome, Lyme disease, mold toxicity, autism, Alzheimer's disease, and Parkinson's disease (among others), is inflammation. Your immune system was just doing its job, triggering an inflammatory response with the intention of curing you, when somehow it got stuck in a loop that it did not know how to break. The response that was initially helpful got out of control. In the same way in which the cell danger response works on a cellular level, the alarm has gone off on an organ system level, and until that alarm is turned off, the body does not know what to do except to keep reacting with an ongoing process of inflammation.

Keep in mind that most of the time, once the invading threat has been dealt with, the body recognizes that the threat is gone and things settle down again. Once the virus is under control or the toxin has been cleared, you usually get well again.

A physician's number-one task, therefore, is to identify exactly where the threat is coming from and to treat that threat. Unfortunately, as many patients have learned, most physicians are

> " A physician's number-one task is to identify exactly where the threat is coming from and to treat that threat. "

not knowledgeable about some of the common major threats that are covered in Part 2 of this book. If the threat isn't diagnosed and treated, it will persist. This is especially problematic with the epidemic illnesses of Lyme disease and mold toxicity. Both result in a persistent inflammatory process that rarely goes away on its own without proper treatment. The newly coined term *chronic environmentally acquired illness (CEAI)* is designed to bring extra attention to this crucial concept.

While most of the time the body returns to normal functioning once the threat has been identified and eliminated, all too often the body still does not recognize that the threat has been removed, and the inflammation persists. This is what I am referring to when I talk about rebooting the immune system. I will address this rebooting process in two sections: first I will talk about rebooting in the presence of an infectious agent, and then I will discuss rebooting in the presence of toxins.

Find the Infectious Agent and Treat It

Of the infections that most often go unrecognized, Lyme disease (with its co-infections) leads the list. However, chronic viral infections and parasites, along with *Chlamydia* and *Mycoplasma*, are similarly difficult to diagnose, and none of these important diagnoses is often considered in conventional medical settings. The prevalence of *Candida* and other fungal infections is rarely recognized, either.

Outlining the specific details of treatment for all of these infections is beyond the scope of this book. The take-home message, however, is of vital importance: When faced with a persistent chronic illness that is inflammatory in nature, searching for a possible infectious cause is critical. Once found, that cause must be treated properly. After this has been accomplished, the inflammatory response often can be turned off, or "rebooted."

We have many approaches to rebooting, most of which are well known to both practitioners and patients. We have antibiotics, antifungals, antiparasitic medications, and antiviral medications. We have a wide assortment of herbal approaches (I recommend all of Stephen Buhner's books, which address these in detail) as well as homeopathic remedies.

Therefore, let me focus on some lesser-known treatments that can be useful as rebooting strategies for infections when the standard medications and supplements are not doing the job adequately. All of these treatments can be used as adjuncts with the more standard treatments that focus on antimicrobial medications; in my experience, doing so makes the conventional approach even more effective.

UVB (Ultraviolet B)

This unique treatment was utilized before the advent of antibiotics but fell into disuse after penicillin was discovered. A modest amount of blood is taken from a vein in the arm and passed through a device that exposes the blood to UVB radiation, and then the blood is cycled back into the body. UVB treatment has been found to be helpful in treating a wide variety of infections. As microbes become less and less responsive to antibiotics, this technology and some of the others noted below are likely to become more and more valuable. How exactly this works is not clear, but the efficacy of this treatment is well documented.

Hydrosol Silver

Often inaccurately called "colloidal silver," hydrosol silver can be used orally, nasally, or intravenously to treat a wide variety of infections. It has become an important integrative component to Lyme and mold treatment for many practitioners. Hydrosol silver by itself can be effective as an antibiotic and an antifungal, and research shows that it is synergistic with those agents as well, making them more effective.

Ozone

Like hydrosol silver, ozone is being used in an increasing number of ways for rebooting the immune response. It can be used intravenously, topically, or with nasal, ear, and rectal insufflation and adds greatly to the treatment options for mold toxicity and Lyme disease.

Hyperbaric Oxygen Therapy (HBOT)

Long used for treating divers with decompression illness, hyperbaric chambers also are used in conventional medicine for treating serious infections that are not responding to the usual antibiotic approaches and for healing wounds that are not healing properly. We are increasingly finding HBOT to be of value in helping to turn off the inflammatory response, and physicians are using it to treat Lyme disease and autism.

Low Dose Immunotherapy (LDI)

Developed by Ty Vincent, MD, as an outgrowth of treating the immune system with LDA (low dose allergen therapy), LDI apparently has the unique ability to reboot the T-regulatory cells that are stuck in an immune-reactive state. Over time, it can cure a wide variety of allergies, and specific treatments are now available for Lyme disease and yeast infections. LDI has shown to reboot, or remove, the inflammatory component of these illnesses so that patients not only feel better but also become much more responsive to treatments that were difficult for them to tolerate while they were in an inflamed state. Courses for physicians in this important modality are taught by the American Academy of Environmental Medicine. Key to the use of LDI is finding exactly the right dosage of material, which can be taken sublingually (under the tongue) or by intradermal injection.

An excellent example is a forty-year-old woman whom I had treated fairly successfully for mold, Lyme, and *Bartonella.* She had reported that she was 80 percent better and had basically gotten her life back, but then had a sudden exacerbation of intense joint pain "all over." She had gone off antibiotics before the attack, but when she resumed them, she did not bounce back as she had in the past. I provided her by injection with her first dose of LDI using Lyme antigens, and within two days, her joint pain resolved almost completely. She has continued LDI treatments with continuous benefit.

Intravenous Immunoglobulin (IVIG)

Specifically used to strengthen the immune system when testing shows low levels of immunoglobulins, IVIG treatment has been of particular benefit in treating PANDAS and PANS. Unfortunately, it is expensive and not often covered by insurance. Several physicians who treat Lyme disease have found it to be of great benefit in improving immune function and speeding up the healing process.

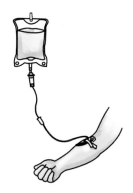

GcMAF (Globulin Component Macrophage Activating Factor)

One of the many ways in which a virus can weaken the immune system such that viral infections are harder to eradicate is by making an enzyme called *nagalase*. Nagalase cleaves the vitamin D receptors from monocytes. This is important because vitamin D must bind to those receptors in order to convert the monocytes to macrophages, which are the cells that specifically go after viruses. When nagalase removes the vitamin D receptor from a monocyte, this component of immune functioning is essentially disabled. This is one of several methods that a virus uses to interfere with the optimal functioning of the immune system and prevent the body from adequately fighting the virus, allowing some viral infections to become chronic. To work around it, we have found that GcMAF can directly activate the macrophages to do their job properly and regain immune control over viral infections, even those that are chronic.

Until recently, GcMAF was available for injection or sublingual use, but currently it is difficult to obtain. An alternative is the probiotic BRAVO, which is prepared in such a way that the GcMAF is concentrated. These materials have been found to be very helpful in treating chronic viral illnesses, and we can monitor their effectiveness by measuring nagalase levels in the body. When nagalase levels have normalized, this indicates that we have accomplished our mission and can discontinue treatment. Because BRAVO is a highly fermented product, patients with mast cell activation may need to be cautious about being exposed to histamine.

Find the Toxin and Treat It

While our world is loaded with thousands of chemicals that have not been evaluated for their safety when it comes to human exposure, the three most common toxins that contribute to chronic illness appear to be molds, heavy metals (especially mercury), and pesticides (especially glyphosate, which is found in the herbicide Roundup; testing for this chemical is newly available from the Great Plains Laboratory, and we are finding, to our dismay, that the vast majority of our patients have high levels of glyphosate in their blood). We have become increasingly aware that the antibiotics in the fluoroquinolones family, including Cipro and Levaquin, cause a type of toxicity centered around chronic, generalized tendonitis that can be difficult to treat. It is beyond the scope of this book to discuss all of these toxins in detail, but I would like to review the current knowledge of mold toxicity and heavy metal toxicity here. Please refer to Chapter 3 on mold toxicity for details on that particular exposure.

Mold Toxicity

Mold exposure includes not only immediate and direct exposure to mold species, but also exposure to mycotoxins made by those mold species, as well as to volatile organic compounds (VOCs), mold fragments, and a wide variety of bacteria, such as *Actinomyces* and *Mycobacteria,* and other irritating substances, such as mannans, proteinases, beta glucans, and hemolysins. This hyperreactive "soup" triggers a well-orchestrated series of biochemical events, which is known as the Biotoxin Pathway and is described in detail in Chapter 3.

The reason I spend so much time on this particular diagnosis is that it appears to be the single most common trigger for persistent inflammatory illness, mast cell activation, porphyria, and pyroluria, all of which contribute greatly to making patients ever more sensitive to their environments and less able to process toxins. Once the mold toxicity has been successfully treated, we see a marked decrease in all of these ancillary conditions, including multiple chemical sensitivities. In the year following successful treatment of and recovery from mold exposure, these conditions go from disabling to minor or even completely resolved. To treat the conditions that are triggered without treating the cause (which in this case most often is mold toxicity) is not a good therapeutic strategy. To reiterate (and

I can't stress this enough), we must always look for and treat *the cause of the underlying illness.*

Heavy Metal Toxicity

In our modern environment, we are exposed to more mercury, lead, arsenic, cadmium, nickel, and aluminum than most people realize. A full discussion of all of these potentially toxic metals is beyond the scope of this book, but I would like to focus on mercury as one of the most common and debilitating of these exposures.

The main symptoms of chronic mercury toxicity include cognitive dysfunction, headaches, fatigue, and paresthesias. These symptoms are common to all of the inflammatory conditions I have been examining, so it can be difficult to tease out mercury toxicity as a contributing diagnosis unless we look for it. Also important is that even minuscule amounts of mercury can poison the key enzyme of methylation, methionine synthase, which works with methyl cobalamin (vitamin B12) to convert homocysteine to methionine. This severely compromises our ability to methylate (see Chapter 14), which is important in our ability to detoxify, make energy, repair DNA, and deal with inflammation.

Measuring a person's mercury load can be difficult. Merely doing a hair analysis has been shown to be inaccurate. Like other heavy metals, mercury usually binds tightly to certain body tissues, and we have to use chelating (binding) agents to pull the heavy metal out of the body so that it can be accurately measured. For example, mercury has an affinity for binding to nerve and brain tissue, whereas lead has an affinity for binding to bone.

Most experts consider the DMPS challenge test to be the most accurate way to measure mercury levels, and by combining the correct dosages of DMPS and EDTA (which are chelating agents), we can optimally bind most of the heavy metals. To measure heavy metals accurately, DMPS and EDTA are administered intravenously over thirty minutes, and then the patient's urine is collected over the next six to twenty-four hours and sent to a laboratory that specializes in measuring heavy metal toxicity.

For very sensitive patients, however, this intravenous infusion can be difficult to tolerate. The Mercury Tri Test, developed by Dr. Chris Shade and available through Quicksilver Scientific, combines analyses of blood, urine, and hair to derive an assessment of mercury load (but not other heavy metals).

Once we learn that heavy metal toxicity is a part of the bigger picture, we need to carefully assess whether a sensitive or toxic patient can tolerate the wide variety of treatments used to remove that burden. At times, we have to wait until the patient is stronger and less reactive before chelation therapies can be started.

The Role of the Gut in Immune Functioning

While I have devoted a whole chapter to rebooting the gastrointestinal system, I would like to emphasize the interrelationship between the gut and the immune system. Approximately 60 percent of the immune system is associated with the gastrointestinal system. This portion of the immune system is called the *gut-associated lymphoid tissue,* or GALT. Surrounding the gut are large patches of lymphocytes (important immune cell lines) called *Peyer's patches,* which process potentially harmful microbes. The Peyer's patches produce antibodies to invaders and immunoglobulin A to create a line of major defense.

Of particular relevance is that if an invading microorganism produces toxins or creates an infectious or inflammatory response in the cells of the intestinal lining, it can weaken the tight junctions between intestinal cells, which serve as a key protective barrier to keep organisms and other substances in the gut from getting into the body. The central chemical that regulates the opening and closing of these tight junctions is *zonulin.* Toxins and microbes can interfere with the regulation of zonulin.

Once these cell-to-cell connections have been loosened or opened, toxins and other large molecules can enter the body and do damage. This is also the mechanism by which food allergy occurs; when large proteins (never intended to get through these barriers) seep through these weakened cellular connections, they can get into the bloodstream and stimulate the body to make antibodies to them, creating allergies to those protein molecules. You may have heard this disturbance in the function of tight junctions referred to as *leaky gut.* Leaky gut is common in chronically ill patients and needs to be addressed as a part of treatment. The most common foods that create allergies are gluten and dairy, but corn, sugar, citrus, pork, and the nightshade family of plants (which includes tomatoes, potatoes, eggplant, and bell peppers) can be problematic as well.

Often patients need to eliminate these foods from their diets and then reintroduce them to learn which ones are contributing to their symptoms.

Recently, we have seen renewed interest in the complexity of intestinal function, which is termed the *gut microbiome.* It is now recognized that restoring the integrity of this biome is a key component of treatment. I will discuss this topic in more detail in Chapter 12 on rebooting the gastrointestinal system.

Purinergic Rebooting

As discussed in detail in Chapter 8 on the cell danger response, one of the most fundamental components of chronic inflammation is the continuous stimulation of purinergic receptors by purines such as ATP. Dr. Robert Naviaux's groundbreaking research on the use of the purinergic receptor blocker suramin to facilitate the shift back to normalcy gives us hope that a new class of medications (currently being worked on by the pharmaceutical industry) may help us more easily reboot the immune system.

Polyvagal Rebooting

In Chapter 9, I discussed the importance of the vagus nerve in coordinating a wide range of biological functions. Chronic inflammation, created by the persistent release of cytokines that is a major component of the illnesses that are the subject of this book, has an important effect on the vagus nerve. An inflamed vagus nerve influences gut motility, mast cell activation, respiratory function, and cardiac function. This nerve, too, must be rebooted to enable the immune system to restore its normal functioning.

NAET (Nambudripad's Allergy Elimination Techniques)

In 1983, Dr. Devi Nambudripad began using a combination of acupuncture, acupressure, energy measurement and balancing, and kinesiology to treat a wide variety of allergy conditions. I began to encounter ultrasensitive patients in 1991, working in San Diego at the Livingston Clinic. When I ran out of options for treating these sensitive patients, I referred many of them to Dr. Nambudripad in Los

Angeles, where at times she worked miracles. Using her combination of assessments and treatments, she successfully desensitized many of them so that we could resume treatment and complete health could be restored. She has trained thousands of practitioners to continue her work. To learn more, go to www.naet.com. I also recommend her classic book *Say Good-bye to Illness.*

Autoimmunity

Autoimmunity literally means that the immune system is attacking itself. A part of the immune system (which I alluded to previously as the cell line called T-regulatory cells, or "T-regs") somehow gets the idea that certain tissues in the body are so similar to certain chemicals or microbes it has been analyzing that it makes antibodies to those tissues.

One of the best-known examples is that as a body processes a *Streptococcus* infection, it can create breakdown products that look similar to heart muscle. The immune system may then start to make antibodies to heart tissue that take the form of rheumatic fever. One of the main reasons to distinguish strep throat from a viral throat infection is that if the strep infection can be treated with penicillin early on, the creation of those infectious by-products can be prevented, and hence the development of rheumatic fever can be prevented. In fact, rheumatic fever was a common consequence of strep infection until the discovery of penicillin but now is quite rare.

The type of inflammation set off by mold toxins, Lyme disease, and its co-infections is a common cause of this confusion of the immune system in which it misidentifies friend as foe. *Virtually every form of autoimmune illness can be triggered by these processes,* so if someone is diagnosed as having an autoimmune illness, it would make sense that he or she be evaluated for mold toxicity and/or Lyme disease early in the course of illness. We have found that if it is addressed early and the inciting cause of autoimmunity is dealt with properly, the autoimmune illness can be *cured.* If there is a significant delay in recognizing or treating that underlying cause, the autoimmune disease will settle in and be much less amenable to a cure.

Common examples of autoimmune illness include Hashimoto's thyroiditis (the most common cause of hypothyroidism), rheumatoid arthritis, multiple sclerosis, inflammatory bowel disease (IBD), and celiac disease.

Summary

Once the immune system has been engaged in a battle with a microbe or toxic substance, if it does not win that battle quickly, it may get stuck in a persistent inflammatory response that it cannot turn off. We are learning that this response is a key component of most chronic illnesses. The most important treatment strategy for dealing with "stuck" inflammation is to find the infection(s) and/or toxin(s) responsible for this response and treat them thoroughly. This is not a simple task, and it can take months or years. Still, it can be done, and that effort must be the centerpiece of treatment; trying to maneuver around it will not work. Even if the cause(s) of inflammation have been addressed correctly, sometimes the immune system still does not know how to reboot itself, and we need to consider the alternative methods for complete rebooting described in this and the accompanying chapters.

I hope I am making it clear that while this stuck response is indeed complicated, it is something that we are learning how to deal with better and better. With persistence and diligence, there is legitimate hope for all of those patients who have been suffering for many years.

CHAPTER 11

Rebooting the Endocrine System

A Hormonal Balancing Act

The chronic inflammatory illnesses that are the subject of this book have widespread effects on the body. Because inflammation is the primary event, it is somewhat obvious that joint and muscle pain and fatigue would be the usual symptoms. It is less obvious that hormones would be affected, but in fact this happens in the majority of patients who are suffering from these illnesses. The endocrine system refers to hormones, which are chemical messengers made by our glands that regulate major functions within our bodies: metabolism, responses to stress, and appetite, among others.

The key concept here is that the inflammation that is provoked by toxins, such as mold, and infectious agents, such as Lyme and *Bartonella,* also inflame areas of the brain such as the pituitary gland, which is located at the base of the brain. The pituitary gland is connected to the hypothalamus, which neurologically helps regulate the pituitary gland's production of important materials such as adrenal hormones, thyroid hormones, sex hormones, and antidiuretic hormone (ADH). Equally important but easily overlooked is the fact that our hormones are intimately interrelated in the way that they influence each other. One dysregulated hormone can have a profound influence on all the others.

What we commonly see clinically is that chronically ill patients produce inadequate adrenal, thyroid, and sex hormones, which adds to and complicates their illness. When I first began to work with myalgic encephalopathy/chronic fatigue syndrome and fibromyalgia in the late 1980s and early 1990s, I found these hormonal deficiencies in the majority of my patients, and I was encouraged that simply treating these conditions improved the majority of them. Jacob Teitelbaum, MD, with his SHINE protocol, was able to demonstrate that when these deficiencies, along with magnesium deficiency, food allergies, and dysbiosis, are treated, fibromyalgia is an eminently treatable condition.

At the same time, conventional medicine was struggling to define these conditions and offering little, if anything, in the way of diagnosis or treatment. Unfortunately, this led to patients' illnesses being labeled "psychosomatic," and many physicians viewed ME/CFS as a psychological illness rather than a physical one. It was not until the arrival of the nerve pain medication Lyrica (pregabalin) in the mid-2000s that conventional medicine felt it had a treatment, and suddenly fibromyalgia became a "real" diagnosis. Unfortunately, Lyrica does not cure fibromyalgia, but it does relieve some of the symptoms in a moderate percentage of the patients who are treated with it. For those of us who were treating fibromyalgia at that time, its great benefit was in legitimizing fibromyalgia as an illness and taking the psychological stigma away from our long-suffering patients.

Over the years, my patient population began to change. Those patients with ME/CFS and fibromyalgia who were readily treated and cured simply by addressing hormone balance, magnesium levels, dysbiosis, and food allergies were fewer and fewer, and the patients who presented to me were more profoundly ill. Perhaps this was caused partly by my increasing specialization in these illnesses, whereby patients who were not responding to previously effective protocols were referred to me by other health professionals. I also believe that we were beginning to see the tip of the epidemics of Lyme disease and mold toxicity—those more recent patients were indeed more compromised.

The straightforward treatment of endocrine issues that had worked so quickly and effectively in the past now worked less well. It helped, but only a little, and we had to turn our attention to working out how to diagnose these evolving illnesses so that we could treat them properly. We are still working on it. Because this area of environmentally acquired illness is not well recognized by conventional medicine, very little research is dedicated to it, despite the CDC's acknowledgment in August 2013 that it was seeing 300,000 new cases of Lyme disease every year.

The understanding of the epidemic of mold toxicity lags far behind that of Lyme. With very little research or funding available, those who are pioneering the treatment of these illnesses have been wrestling with the inertia of conventional medicine and our primitive understandings of how to treat these illnesses. Despite these limitations, we have made an excellent beginning.

What I am alluding to here is that our past recognition of the importance of endocrine disturbances and the role they play in chronic illness has taken a back seat to the treatment of mold toxicity and Lyme disease, and I hope to bring this back into awareness.

The HPA Axis (Hypothalamic/ Pituitary/Adrenal Axis)

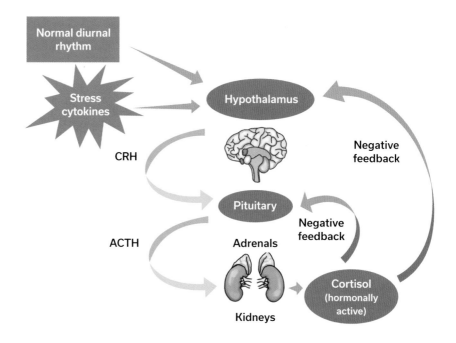

Figure 11.1. The HPA axis is thought of as the body's central stress response system.

As shown in Figure 11.1, the HPA axis is a complex interaction between the hypothalamus, the pituitary gland, and the adrenal glands. It is usually thought of as the body's central stress response system, and it is predictably underactive in the presence of most chronic illnesses in which fatigue is a central symptom.

The hypothalamus sends signals to the two separate portions of the pituitary gland, the anterior and posterior pituitary, to either stimulate or inhibit the production of hormones.

The anterior pituitary produces:

- **ACTH (adrenocorticotropic hormone),** which stimulates the adrenal glands to produce their hormones. Some of these hormones are clearly related to stress, such as cortisol, but others regulate energy and the production of other hormones, such as DHEA (dehydroepiandrosterone) and the mineralocorticoids, which help regulate blood pressure.
- **Growth hormone,** which stimulates growth, cell reproduction, and regeneration.
- **FSH (follicle-stimulating hormone),** which stimulates the ovaries or testes to make estrogens, progesterone, and testosterone.
- **LH (luteinizing hormone),** which works with FSH to regulate ovarian or testicular hormones.
- **TSH (thyroid-stimulating hormone),** which regulates the production of thyroid hormones.
- **Prolactin,** which stimulates the production of breast milk in women but also plays a role in immune regulation and is partly controlled by VIP (see page 55).

The posterior pituitary produces:

- **Oxytocin,** which stimulates breast milk production and causes the uterus to contract after childbirth. It also has important regulatory effects on empathy, generosity, and orgasm and modulates fear and anxiety.
- **ADH (antidiuretic hormone),** which stimulates the kidneys to increase water reabsorption.

Even at a glance, you can see how complicated the HPA axis is and how many hormonal interactions are possible. The main point I wish to impart is that chronic illnesses, with their immediate effects on the HPA axis created by the inflammatory response that they incite in the brain (which I covered in Chapter 3), have the potential to markedly interfere with hormonal production and balance—and they do. Both mold toxicity and Lyme disease directly interfere with the production and regulation of these hormones, so we can anticipate that the vast majority of chronically ill patients will have some difficulties with hormonal balance.

Let's look at each of these hormones separately, with an emphasis on the hormones that are most significantly affected by these illnesses. I will focus on the specific symptoms that each of these hormonal deficiencies can create and how those symptoms can be rebooted or treated.

Even though the medical evaluation and treatment of hormones is covered by the specialty field of endocrinology, for reasons that are unclear to me, endocrinologists have confined themselves to the treatment of severe hormonal imbalances caused by damage to the organs that produce these hormones. Conditions in which the body temporarily produces less hormone than it needs, creating "mild" deficiencies, are not considered under their domain. Therefore, do not be surprised if your endocrinologist does not take into consideration the hormonal deficiencies described on the following pages. These deficiencies appear to be better appreciated, and treated, within the growing field of functional medicine.

If you think any of these descriptions might apply to you, I suggest that you look for a functional medicine physician to assist you with diagnosis and treatment. The organizations that have trained physicians in functional medicine include the Institute for Functional Medicine, the International Society for Environmentally Acquired Illness, the International Lyme and Associated Diseases Society, the Forum for Integrated Medicine, the Academy of Integrative Health & Medicine, the American College for the Advancement of Medicine, and the American Academy of Environmental Medicine.

Adrenal Hormone Deficiencies

The adrenal glands are two small cone-shaped organs perched atop each kidney. They are primarily responsible for how the body responds to stress. These glands do not distinguish between the causes of stress, so physical stressors such as surgery, injury, childbirth, or chronic pain and illness place just as much strain on the adrenal glands as mental, emotional, or spiritual stressors. Feeling bad on a daily basis, for any reason, is more than sufficient to overtax the adrenal glands and deplete their production of hormones over time.

To oversimplify a little, the adrenal glands make three major types of hormones: DHEA, cortisol, and mineralocorticoids. A person can have a deficiency of one, two, or all three of these adrenal hormones. Because these deficiencies are relatively easy to test for, diagnose, and treat, the initial evaluation for all patients with chronic illness should include these measurements of adrenal function. While correcting these deficiencies may not cure the underlying condition that has created them, patients often feel much better and have more energy to pursue definitive treatments.

DHEA

The largest amount of hormone produced by the adrenal glands is in the form of DHEA, which is a precursor to the important hormones estrogen, progesterone, and testosterone, as well as being important in and of itself. Symptoms of DHEA deficiency include fatigue, tiredness, exhaustion, cognitive impairment, depression, decreased libido, a risk of recurrent infections, and a general sense that "I just don't feel right."

DHEA is relatively easy to measure. Most labs measure a form called DHEA-S or DHEA sulfate, which in my opinion is not as accurate to measure or track as "plain" DHEA, which is referred to as *unconjugated DHEA*. However DHEA is measured, the results must be correlated to the patient's age to be meaningful. Males around the age of twenty should have high levels of unconjugated DHEA, around 1,200 ng/dL. By age ninety, the anticipated level is 180 ng/dL. A thirty-year-old male with a DHEA level of 250 ng/dL would technically fall "within the normal range" of 180 to 1,200. If you did not understand the age-related nature of these results, you would be missing the fact that this young man had a significantly low DHEA level (for his age, it should be around 1,000), which would lead you to miss the diagnosis of adrenal deficiency and the opportunity to treat it.

Because stress is an underlying feature of all chronic illness, it should come as no surprise that the vast majority of patients who have been ill for a long time have low DHEA levels and likely would benefit from treatment. While there are a few contraindications for using DHEA (including active prostate or breast cancer), it is otherwise quite safe and very helpful to supplement adrenal glands that cannot keep up with the needs of a stressed-out system. As always, all patients should be diagnosed and treated only by knowledgeable healthcare providers.

Cortisol

Cortisol is an adrenal hormone that assists the body in coping with stress. Symptoms of cortisol deficiency include fatigue, exhaustion, a tendency toward allergies, abnormal hair growth, and a tendency to miscarry.

Cortisol is typically made in a diurnal pattern, meaning that it is manufactured primarily during sleep. Levels peak at about eight in the morning and gradually drop throughout the day, which means that a single measurement of blood cortisol may not reflect the full picture.

In the past, when it was less expensive and more readily available, I found the cortrosyn stimulation test to be the most accurate for measuring cortisol levels. For quite a few years now, however, I have found that measuring cortisol in saliva, done multiple times throughout the day, is a better and easier method to determine how the adrenal glands are producing cortisol.

If testing determines a low cortisol level, it can be supplemented with the pharmaceutical agent hydrocortisone acetate. In small doses, this supplement has been quite effective at improving energy levels and immune function in compromised patients. William Jefferies, MD, an endocrinologist, wrote the classic book *Safe Uses of Cortisol,* which provides excellent guidelines about how to prescribe and use this medication safely and effectively.

Mineralocorticoids

A third group of hormones made by the adrenal glands are the mineralocorticoids. These hormones help regulate blood pressure, raising it when it is low. Symptoms of mineralocorticoid deficiency include low blood pressure with pronounced dizziness or light-headedness upon standing or sitting up, referred to as *orthostatic hypotension.* Other possible symptoms are fatigue, tiredness, and exhaustion.

Think of it this way: If your home had low water pressure and you took a shower, the water would come out as a trickle. Low blood pressure is like that; that your blood does not adequately perfuse your body. With blood flow that is not quite sufficient, it is easy to see how you would be predisposed to fatigue.

I encourage patients to obtain a series of blood pressure readings. If they routinely run 110/70 mm Hg or less and the patient reports dizziness and/or light-headedness when he or she stands or sits up, a trial of the medication fludrocortisone (Florinef) or midodrine should be considered. It is also a good idea to measure blood pressure both after lying flat for five minutes and then sitting up. A drop in blood pressure of 10 to 20 mm Hg would confirm the diagnosis of orthostatic hypotension.

Thyroid Hormone Deficiencies

The thyroid is a butterfly-shaped gland that wraps around the front of the neck. The essential function of thyroid hormones is the regulation of metabolism.

Deficiencies in thyroid hormone, termed *hypothyroidism,* create a sluggish overall chemistry that affects every part of the body. Symptoms include fatigue, tiredness, exhaustion, constipation, temperature dysregulation (usually low body temperature, or feeling cold all the time), hair loss, dry skin, cognitive impairment, menstrual abnormalities, weight gain, and loss of hair on the outer third of the eyebrows. Conversely, excesses of thyroid hormone, called *hyperthyroidism,* speed up metabolism, creating the classic symptoms of palpitations, increased heart rate, elevated blood pressure, and anxiety. The majority of chronically ill patients have low thyroid function, so that is what I will focus on here.

The first difficulty in diagnosing hypothyroidism is in obtaining the proper testing. For decades, patients have arrived at my doorstep certain that they have low thyroid, even though their family physicians, internists, and/or endocrinologists have repeatedly told them that they do not, based on testing. The standard test for hypothyroidism has long been the measurement of TSH, the pituitary hormone that regulates the production of thyroid hormones from the thyroid gland. Many physicians are under the impression that this test alone can determine who has a low thyroid.

Unfortunately, as our clinical experience accumulates, we have come to realize that TSH does not give a complete picture of thyroid production and balance. To assess this system accurately, we need to measure not only TSH but also T3 (triiodothyronine) and T4 (thyroxine), the two main hormones produced by the thyroid gland. In addition, in many patients with subtle but clinically significant deficiencies of thyroid hormone, we need to measure reverse T3 (RT3) to get a full picture of their thyroid function. With this more complete overview, we have discovered that many patients who have been told that their thyroid function is "normal" do indeed have thyroid dysfunction and improve with the proper treatment.

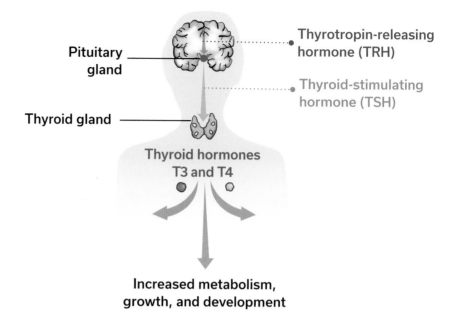

Pituitary gland

Thyrotropin-releasing hormone (TRH)

Thyroid-stimulating hormone (TSH)

Thyroid gland

Thyroid hormones
T3 and T4

Increased metabolism,
growth, and development

Figure 11.2. The thyroid gland produces thyroid hormones, which regulate metabolism.

Here is how the production of thyroid hormones works: The thyroid receives a message from the pituitary gland in the form of TSH, which stimulates the thyroid to make more thyroid hormones. First, the thyroid produces T4, the inactive form of thyroid hormone, which must be broken down (by the removal of an iodine molecule) into T3, the active form. When this system is not working well, the thyroid produces RT3, which inhibits the normal conversion of T4 to T3. Over time, RT3 accumulates and further prevents this conversion. It is not uncommon for a patient to present with the classic signs of thyroid deficiency with a normal TSH level and even a normal T4 level, but low T3 and high RT3. "Standard" testing will suggest that this patient has a properly working thyroid gland, but when we look further, we often find that the patient's body is not working properly and requires thyroid supplementation to resume normal functioning.

The major thyroid hormone replacement provided by family physicians and endocrinologists is T4 (levothyroxine, or Synthroid). While this works beautifully for most patients, many physicians are unaware that for certain other patients, this form of treatment will only cause further buildup of RT3 and make it even more difficult for their bodies to convert T4 to T3. For these patients, using T3 in the form of liothyronine sodium (Cytomel), compounded sustained release of T3, or Armour thyroid (a combination of 20 percent T3 and 80 percent T4) is needed to provide the proper hormone replacement.

Over time, the use of T3 will take the pressure off the buildup of RT3, which will normalize, and normal thyroid hormonal conversion can prevail. I personally have treated several thousand patients in this manner with excellent results, so I am convinced that this is the future of how the thyroid gland must be evaluated and treated.

As you can see, this form of treatment is a kind of "rebooting" to get the thyroid gland back to normalcy. The diagnosis was first popularized by Dennis Wilson, MD, and sometimes is referred to as Wilson's syndrome (not to be confused with the rare copper accumulation disorder called Wilson's disease). To evaluate and treat patients in this manner is not yet common practice in conventional medicine, so people who seek this approach will need to find physicians who are trained in functional medicine.

To provide additional perspective here, the most common cause of hypothyroidism is Hashimoto's thyroiditis. This is an autoimmune condition that begins with inflammation of the thyroid gland. At the outset it often produces symptoms of hyperthyroidism, but as the disease progresses, the thyroid gland burns itself out and becomes hypothyroid. The inflammatory immune dysregulation produced by mold toxicity and Lyme disease is a frequent cause of autoimmunity, so anyone with an autoimmune process should consider seeking an evaluation for those conditions. We have found that, if caught early, the autoimmune process may be reversible. If it persists, reversal becomes less likely.

You can see that because it both predisposes people to autoimmune illness and creates dysregulation of the HPA axis, chronic inflammatory illness may need to be addressed. Conversely, by rebalancing the thyroid gland, we can help patients improve their functioning while addressing the underlying causes of their illness.

Sex Hormone Deficiencies

Estrogens

The ovaries make a variety of estrogens that have different physiological functions, even though low levels of estrogens are often referred to simply as "estrogen deficiency," as if estrogen were a singular entity. These include estrone (E1), estradiol (E2), and estriol (E3). Symptoms of estrogen deficiency include hot flashes, night sweats, insomnia, mood swings, depression, anxiety, cognitive impairment, fatigue, vaginal dryness, and decreased libido.

It is important to understand that the brain and vaginal area have estrogen receptors, and when those receptors do not receive the stimuli they are used to receiving (as a result of having the proper amounts of estrogens in the body), they respond by letting the body know that they are unhappy. The heart also has estrogen receptors, and I see a few female patients each year who have palpitations or cardiac arrhythmias that resolve when these women are given the proper dose of estrogens. Women who have suffered from migraine headaches sometimes discover that without adequate estrogens, they have recurrences or exacerbations of migraines until the proper hormone balance is achieved.

Some women have come to view the passage through menopause as a badge of achievement and, especially if they have had little difficulty with this passage themselves, encourage all women to be strong and fight their way through the symptoms for many years. As a man, I may not be entitled to comment on this matter, but as a doctor, I will express my opinions here. In the same way that thyroid deficiency clearly represents a physical *need* for more hormone so that the body can function properly and hence carries no stigma, I think estrogen deficiency should be considered a *need* as well. It is no minor issue for a woman to be unable to sleep, swing through intense emotional highs and lows, and experience fatigue, cognitive impairment, and vaginal dryness. These symptoms can have serious repercussions on her well-being and health and affect her family and friends as well. Treating estrogen deficiency is not, as some believe, a "luxury" but rather a biological necessity, and I hope that I am conveying the importance of addressing it in a responsible manner.

When a woman starts going through menopause, the first thing that happens is that her ovaries begin to make less progesterone. This leads to an imbalance in which she has relatively more estrogen than progesterone, called *estrogen dominance*. Providing additional progesterone at this point can stabilize the balance of hormones. After a while, the ovaries start to make less and less estrogen(s), which initiates full-blown menopausal symptoms. For some women, these symptoms are minor and do not cause much difficulty, but others experience a profound disturbance.

Unfortunately, when it became clear that synthetic horse urine–derived estrogens caused a marked increase in breast cancer, many women assumed that *all* estrogens were bad. Understandably unwilling to risk breast cancer, women became afraid to take any estrogen, and many have suffered greatly as a result. Taking bioidentical or "natural" estrogens does not appear to subject women to these same risks, and many would benefit from their judicious use.

Although one might think that this discussion is confined to women in their mid-forties and older, I am seeing younger and younger women—some in their twenties—with clear symptoms of estrogen deficiency that is confirmed by laboratory testing. These women respond to taking bioidentical hormones with immediate improvement. I think we need to be aware that sex hormone deficiencies are more prevalent in younger patients than we thought.

I recommend that any woman contemplating hormonal treatment begin by measuring all three estrogens—estrone (E1), estradiol (E2), and estriol (E3)—as well as progesterone, testosterone, and DHEA to understand the bigger picture. I have found the Genova Hormonal Assay to be the most accurate for determining a patient's hormonal needs. Another useful measurement to include in this evaluation is sex hormone binding globulin (SHBG), a protein that, as its name suggests, binds to estrogen and testosterone in the blood. If SHBG is elevated, even if blood tests reveal normal hormonal levels, these hormones may be so tightly bound to this substance that they are not available to the body as they normally would be. The body thus may behave as if it were deficient in these hormones because the hormones present in their bodies are not available for use.

Particularly relevant to our discussion of chronic inflammatory illness, mold toxicity and Lyme disease directly affect pituitary function and are common causes of these hormonal imbalances. Treating the underlying cause may restore normal hormonal balance, but sometimes the system needs to be "rebooted" by giving the correct hormones and reminding the body of the balance that it requires. There is a complex, interwoven relationship between estrogens and thyroid hormones, and hormonal dosages may need to be adjusted to provide balance. Both sex hormones and thyroid hormones may also be influenced by adrenal hormones, so all of these hormonal factors need to be taken into consideration to help the body regulate itself properly. All prescribed treatments should, of course, be administered only by trained healthcare providers.

Progesterone

While estrogen receives the lion's share of attention as the most important of a woman's sex hormones, it functions in close relationship to progesterone. Many women who have undergone hysterectomies are offered estrogen replacement alone, with the mistaken concept that because a woman no longer has a uterus, we don't need to worry about helping to slough the lining of that uterus every month with her menstrual period. Removing the built-up uterine lining each month is indeed an important function of progesterone, but thinking that this is progesterone's only job would be a shortsighted understanding of the role that progesterone plays in health.

Progesterone is an important precursor to other hormones, especially adrenal hormones. It helps calm or quiet the nervous system (a deficiency of progesterone can contribute to anxiety and insomnia); stimulates new bone growth (which is important for women who are at risk for osteoporosis); improves libido; and, especially important, regulates the sensitivity of estrogen receptors. After a hysterectomy, if progesterone levels are low (which is almost universal), progesterone should be a part of the hormone replacement program. For all women contemplating bioidentical hormone replacement, estrogen and progesterone should be used together in the proper balance.

Testosterone

Because testosterone is the quintessential male sex hormone, I will focus this discussion on testosterone deficiencies in men. Please keep in mind that women also can be deficient in this hormone and may need to take small doses to restore optimal health. Men also make estrogen and progesterone, but in far smaller concentrations than women. Once again, the balance of these hormones is what is important.

Symptoms of testosterone deficiency include fatigue, tiredness, decreased stamina, low libido, erectile dysfunction, muscle weakness, depression, mood swings, hot flashes, palpitations, insomnia, an inability to concentrate, and antisocial tendencies.

As men age, they are prone to the equivalent of menopause, appropriately referred to as "male menopause." If a man has any of these symptoms of testosterone deficiency, testosterone levels should be measured to determine how to approach treatment. Not only do we need to know the amount of testosterone in the blood, but we also need to measure SHBG. As discussed in the section on estrogens, SHBG binds to testosterone. A high level of SHBG prevents adequate testosterone from reaching the tissues, causing low "free" testosterone, which is now being measured more frequently.

Once it is clear that a testosterone deficiency exists, the age of the patient needs to be kept in mind. Taking testosterone in the form of a cream, patch, or injection can send a message to the brain that will turn off testosterone production. The body gets the message that it has all the testosterone it needs, so it doesn't need to bother making any more. In younger men, this may cause a prolonged turnoff of testosterone production, which can be difficult to bring back online. The medication clomiphene citrate (Clomid) is often used to stimulate the production of luteinizing hormone (LH) in younger men, which in turn stimulates the testes to make testosterone. In this way, we do not engage the negative biofeedback loop that occurs when taking testosterone turns off the body's production of this important hormone.

Another important factor to consider when evaluating testosterone levels is estrogen levels. As men become obese, fat cells called *adipocytes* begin to produce more of an enzyme called *aromatase*. Aromatase converts testosterone to estrogen, which can alter the hormonal balance and affect all the functions of testosterone. Small doses of aromatase inhibitors can prevent this conversion, lowering the levels of estrogens and making testosterone more available.

It is clear that testosterone is important in men's health, and not just for the treatment of decreased libido or erectile dysfunction. Unfortunately, some athletes have abused testosterone and its precursors, and the FDA has made testosterone a scheduled drug, placing it in the same class as narcotics. This has made it somewhat more difficult for physicians to prescribe testosterone, but that does not diminish the need for those who are medically deficient.

Pituitary Hormone Deficiencies

ADH (Antidiuretic Hormone or Vasopressin)

Early in his work with mold toxicity, Dr. Ritchie Shoemaker found that many of his patients complained of frequent urination and insatiable thirst. When tested, many of them had a low level of ADH in the blood and a low serum osmolality (a low concentration of sodium and other electrolytes). It appears that mold toxicity causes these deficiencies in some patients. They can be addressed with small doses of desmopressin, which is available by prescription in tablet, nasal spray, or injection form. Again, while this treatment will not cure the mold toxicity, it can improve patients' physiological functioning while they work on the underlying causes.

Parathyroid Hormone (PTH or Parathormone)

Parathyroid hormone is not directly affected by the HPA axis, but I want to bring an awareness of this condition to my readers because I believe that it is underdiagnosed and undertreated.

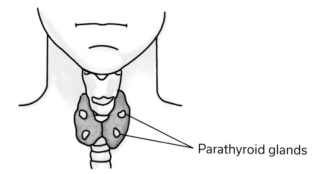

Figure 11.3. The parathyroid glands as seen from the back of the thyroid gland.

Parathyroid glands

The parathyroids are four tiny round glands situated behind the thyroid (see Figure 11.3). They regulate the concentration of calcium in the bloodstream. It is interesting that the body believes this task to be so important that it designates a gland for this purpose. No other mineral gleans so much individual attention.

Hyperparathyroidism is a condition in which one or more of these tiny glands has become enlarged, has become a tumor, and is making too much PTH. Symptoms of hyperparathyroidism include chronic fatigue, muscle aches and pains, bone pain, insomnia, poor memory and/or concentration, depression, and headaches. Left untreated, it can lead to osteoporosis, kidney stones, hypertension, cardiac arrhythmias, or kidney failure. Note that the initial symptoms of hyperparathyroidism easily can be (and often are) mistaken for myalgic encephalomyelitis/chronic fatigue syndrome or fibromyalgia. We have learned in recent years that a calcium level greater than 10 mg/dL (as measured by a routine blood test) may be an early indication of hyperparathyroidism. Once this diagnosis is suspected, a blood test for PTH should be done. The "normal" level for PTH is listed as 10 to 65 pg/mL, but a host of recent publications show that any level greater than 35 pg/mL is suggestive of hyperparathyroidism.

Until recently, major head and neck surgery was the standard approach to removing hyperparathyroid tumors. In 1993, minimally invasive surgical techniques were demonstrated to safely remove these tumors in just fifteen minutes and revolutionized our options for treatment. Improvement in symptoms can occur within days of receiving surgical care.

Unfortunately, this diagnostic and therapeutic information has not yet been embraced by most physicians, and many patients with elevated but fluctuating calcium levels and "borderline" PTH levels go undiagnosed and untreated for years. For those patients who have been suffering from fatigue, pain, and cognitive impairment, I present this information so that if you have ever had an elevated calcium level, you can seek additional evaluation so that you can receive the proper care if warranted.

Insulin

Insulin is not a pituitary hormone itself, but it interfaces with the pituitary hormones in significant ways. Stress affects the production of insulin, so adrenal and thyroid influences have an impact on how much insulin the body produces and how it is regulated. I've already discussed an excess of insulin, which results in hypoglycemia (see page 177), but a deficiency of insulin is far more common, resulting in either type 1 or type 2 diabetes. Type 1 diabetes (also known as juvenile diabetes) represents a complete failure on the part of the pancreas to make insulin; the more common type 2 diabetes represents either a deficiency of insulin production or an inability to use insulin effectively.

A major component of what is referred to as *metabolic syndrome* (which involves insulin deficiency, elevated blood pressure, elevated cholesterol, and obesity), type 2 diabetes causes blood sugar to rise to levels that can negatively affect health. This rise in blood sugar is referred to as *hyperglycemia*. At first, the pancreas makes extra insulin in an attempt to lower blood sugar levels back into the normal range. Over time, however, the pancreas cannot keep up with the demand, and its production of insulin is not adequate for the task of regulating blood sugar levels. This condition is referred to as *insulin resistance*. With the ever-increasing problem of obesity, estimated to affect 60 percent of the adult population in the United States, insulin resistance has reached epidemic levels.

A complete discussion of insulin resistance is beyond the scope of this book, but I thought it important to remind you that, given the intricate connections among all our hormones, the inflammatory states I have been describing commonly lead to insulin resistance. Measuring blood glucose levels and HbA1C (a measure of average blood sugar levels over a three-month period) should be a part of evaluation and treatment. *Insulin itself has a pro-inflammatory effect, so keeping insulin levels low and steady is helpful in controlling generalized inflammation.* Unfortunately, many chronically ill patients have gained a great deal of weight and, because of leptin metabolism (see Chapter 13), have difficulty losing weight until the inflammatory causes of illness have been properly addressed. Although it adds another level of complexity to diagnosis and treatment, please remember that insulin resistance is treatable.

Summary

All of these hormones (except PTH and insulin) rely on an intact HPA axis to stimulate the pituitary gland to regulate their production. The inflammatory processes that I have been discussing can directly affect these systems and make it more difficult for the body to get back into balance. We are, in essence, trying to hit a moving target. It is important to be aware of how often these hormonal imbalances accompany mold toxicity, Lyme disease, and other causes of chronic inflammation so that we can address them while we are working on what is creating them. Understanding these imbalances allows patients who have been dysregulated and suffering for a long time to experience some degree of immediate relief and improvement.

CHAPTER 12

Rebooting the Gastrointestinal System

Food for Thought

Throughout this part of the book, I have been using "rebooting" as a metaphor for creating a shift in a system that is somehow "stuck." *Homeostasis* is the term used to describe the body's innate ability to right itself when things have gone awry. When mild to moderate stressors affect us, our built-in mechanisms to reverse the effects of those stressors kick in, and we can heal ourselves using those mechanisms without much difficulty. If the effects of those stressors become severe (the definition of which is unique to each person), however, we may not be able to reverse them. This book is devoted to understanding the systems that prevent healing from taking place. It is my hypothesis that if we can understand those systems and their effects, we can strategize how to reverse them.

The rebooting analogy applies most obviously to healing the nervous system, since we are working from one electrical system to another. The gastrointestinal system, it turns out, is considerably more complex (as if the other systems were not complicated enough) because we are dealing with not one system, but many. The major symptoms that patients suffer from when the gastrointestinal system begins to function poorly are:

- Nausea
- Vomiting
- Reflux (heartburn)
- Constipation
- Diarrhea

- Bloating, gas, and distension
- Abdominal pain (which can be localized or diffuse)
- Indigestion

Some of these symptoms are caused primarily by difficulties with *intestinal motility*, which is the ability of the intestines to move food smoothly down the intestinal tract. Motility is controlled by the nervous system, directed mainly by the vagus nerve and the autonomic nervous system. It also is controlled by the muscles that line the intestines, which, by contracting, produce the movement called *peristalsis* that allows food to proceed through the intestinal tract in a steady flow. It is affected by structural elements, too, stuck components of which could include adhesions from an old abdominal surgery or radiation therapy. It also is affected by internal blockage of the intestines, which can occur in constipation, which may not allow food to pass normally as it is being digested.

So intestinal motility, or a lack thereof, is the result of many different systems needing to come together in balance to produce a "normal" bowel movement. Interestingly, medicine has never been able to define what exactly that means. Some experts feel that a bowel movement should occur after every meal because of what is termed the *gastrocolic reflex*, in which food in the stomach stimulates the colon to empty. Others believe that one bowel movement a day is healthy and sufficient. As with many physiological functions, I suspect that each individual will need to define what feels normal and right for him- or herself.

Other symptoms are caused by difficulties with the *digestion* of foods. Again, this process is complicated. We have separate enzymes to digest proteins (which also requires adequate hydrochloric acid in the stomach), carbohydrates, and fats. This means that we need adequate saliva; a functional liver, pancreas, and gallbladder; and adequate stomach acid and digestive enzymes. We also need highly functional *tight junctions*, where every cell that lines the intestines touches the next cell to prevent larger molecules from crossing the intestinal barrier, getting into the circulation, and stimulating the immune system to make antibodies, causing food allergies.

We are now crossing into another system, *intestinal absorption*, in which the membranes that line the intestines must be highly functional to admit those substances that will nourish us and keep out those that could harm us. It is known that 60 percent of the human immune system is present in the GALT (gut-associated lymphoid tissue), which is located just outside the intestinal lining. This system must be functional as well to keep out bacteria, viruses, and parasites. A key line of immune cells in the intestines called *mast cells* have the job of coordinating the response to toxins and infectious agents between the immune system and the nervous system. A high percentage of sensitive patients will develop mast cell activation (see Chapter 5), which makes them extremely reactive to a wide variety of stimuli coming down the gastrointestinal pipeline.

The coordination of these complicated systems also relates to the fact that the intestinal tract produces a larger concentration of neurotransmitters than the brain itself. It is sometimes referred to as the *second brain.*

And then, of course, we have the *gut microbiome,* which has received a great deal of well-deserved attention in recent years. We have learned that interaction between "good bacteria" and possible pathogens is essential to our health and, like everything else I am describing, is way more complicated than was once thought. Again, we are discovering how little we know.

All of these systems must operate in balance, with each affecting the others, in a complicated dance that we call health.

So, when I talk about rebooting the gastrointestinal system, I have to break this subject down into all these different pieces. In order to heal this system once it has broken down, we need to identify which pieces are not working properly so that we can direct therapeutic efforts into the correct areas. Because so many systems are at play here, it should not come as a surprise that many patients develop gastrointestinal symptoms as a consequence of illness. For sensitive patients, these symptoms may be especially debilitating as they discover that their diet is more and more limited as they become increasingly reactive to the foods they eat.

The major causes of chronic inflammatory illnesses, namely the toxins and infections I've been discussing, can contribute to intestinal disturbances in a variety of ways. Mold toxins and *Bartonella* have a direct inflammatory effect on the stomach and intestinal lining and can cause gastric pain, heartburn, diarrhea, and constipation. The intestinal inflammation that they create weakens the tight junctions, allowing the entry of foreign proteins to which the immune system may make antibodies, causing food allergies and autoimmune disease. The disruption of the gut microbiome causes a further inflammatory response. The inflammatory cytokines produced by these interactions can cross the blood-brain barrier and create inflammation in brain tissues, notably the vagus nerve and autonomic nervous

> **All of these systems must operate in balance, with each affecting the others, in a complicated dance that we call health.**

system, which further interferes with intestinal motility and adds to the inflammatory disruption of the intestinal lining. This becomes a vicious cycle.

Treatment of the gastrointestinal tract must be included in a patient's therapeutic plan, as it plays an important role in keeping the inflammatory process going. Simply treating the gastrointestinal system without simultaneously addressing the cause(s) that are central to this disruption will not succeed in moving a patient toward healing. Some practitioners insist that they cannot move patients forward until they address the gut. While they are correct to a certain extent, this is a limited viewpoint that does not begin to incorporate the bigger picture and, as a therapeutic strategy, will not always work.

Evaluating Gastrointestinal Function

Many tests are available to help us understand different pieces of these imbalances in intestinal motility, digestion, and absorption, and more tests are under development. Some are used in conventional medicine, but most are being used solely by functional and integrative medicine practitioners.

SIBO Testing/Stool Testing

In recent years, gastroenterologists have begun to recognize that the overgrowth of a variety of microbes in the intestines causes many of the symptoms I have outlined. Integrative physicians have known this for decades, utilizing laboratories that measure the presence of beneficial and toxic microbes by a test measuring *dysbiosis,* an excellent term that describes this imbalance. A limited form of measuring this imbalance is the testing for small intestinal bacterial overgrowth (SIBO).

The development of commercial assays measuring hydrogen gas and/or methane allows physicians to give patients kits to collect specimens that can help them learn whether certain gases are being made by pathogenic bacteria in the intestines. When positive, these tests can be useful, but more often than not, they fail to identify SIBO, and many patients go untreated. It is difficult to identify the exact intestinal bacteria by stool culture or by PCR technology applied to stool tests, but these methods have improved a great deal in recent years. With a simple stool test, we often can find specific toxic bacteria (such as *Klebsiella*) or yeast (such as *Candida*) or inadequate *Lactobacilli* (the "good bacteria" of the small intestine) and *Bifidobacteria* (the "good bacteria" of the large intestine), which can be supplemented with the appropriate probiotics. Stool also can be tested for the presence of parasites. More common than generally thought, parasites are difficult to find in routine testing unless multiple specimens are submitted to a laboratory that specializes in finding parasites. These tests provide us with information that can be of great benefit in understanding patients' imbalances and how to approach treatment.

Testing for Food Allergies

Many labs can test for food allergies, but in general I do not find these tests to be particularly accurate. Part of the problem is that it is difficult to test for all aspects. There are five major mediators of allergy, called immunoglobulins, and they are labeled IgA, IgD, IgE, IgG, and IgM. These immune materials manufactured by our immune systems react to the presence of substances, usually proteins. Because we are discussing foods, this means that these immunoglobulins set off reactions to foods.

There are two major types of reactions. *Immediate sensitivity reactions,* such as anaphylaxis, which involves swelling of the tongue and lips, difficulty breathing, and hives, occur within fifteen to twenty minutes after exposure to a food. Most commonly, the allergens that cause these reactions are shellfish, strawberries, and peanuts, and they are mediated by IgE.

There are also *delayed sensitivity reactions* mediated by IgG and IgM, which can occur anywhere from an hour or two to seventy-two hours after exposure to a food. Because immediate sensitivities occur so quickly, it is easier to figure out which food is the culprit. With delayed sensitivities, it is often not obvious that a reaction today was caused by a food eaten two days ago. Delayed sensitivities are much more common than is appreciated. Symptoms of delayed sensitivities to foods include joint and muscle pain, fatigue, cognitive difficulties, bronchospasms, allergic rhinitis, eczema, inflammatory bowel disease, irritable bowel syndrome (IBS), psoriasis, urinary frequency and pain, heart palpitations, cholecystitis (inflammation of the gallbladder), enuresis (bedwetting), sinusitis, and autoimmune disease.

The most accurate, gold-standard method for determining food allergies is to do an elimination diet. This consists of eliminating all of the foods that are suspected as possible allergens for seven to ten days (longer periods of elimination may pick up a few additional patients with gluten sensitivity) and then adding them back, in as pure a form as possible, one at a time, every three days. For example, if you are testing dairy, you could test with cottage cheese, sour cream, or milk, but not ice cream, which contains sugar and other ingredients that might compromise the results. If no reaction occurs within seventy-two hours, you can try the next eliminated food. If there is a reaction, you wait three days to clear your intestinal system (it takes about three days for food to pass completely through the gut) and then try the next food.

When testing foods, it is important to eat a *lot* of that food to provoke a reaction. Small amounts may not cause a reaction, and you may miss important information.

The most common foods that provoke delayed sensitivity reactions are wheat (and other forms of gluten), dairy (especially cow dairy, but we are learning that many people react to dairy from sheep and goats as well), sugar, corn, and citrus. There is a lot of cross-reactivity between soy and dairy, so if you have issues with dairy, it may be helpful to eliminate soy as well. Eggs, caffeine, and alcohol also can cause allergic reactions, but to a lesser extent. If you have joint pain, nightshades may be the culprit; eliminating tomatoes, potatoes (but not sweet potatoes), bell peppers, and eggplant may be helpful.

Testing foods with an elimination diet allows us to see the reactivity of all the immunoglobulins. Blood testing typically tests either IgE or IgG and IgM antibodies, but not all three; there is no commercial testing available for IgA or IgD antibodies. What this means is that blood testing is not as comprehensive as we need it to be to determine how the body reacts to a particular food.

I also have found that few of the commercially available tests give me clinically useful information that allows me to help a patient figure out which foods really should be avoided. Often, those tests report foods that the patient should not eat, but when the patient actually eats them, he or she does fine, making me question the accuracy of the results. The best of these tests, in my opinion, is the ELISA/ACT, which gives more specific and accurate information than the others.

Testing for Hypoglycemia

Hypoglycemia is common in chronically ill patients. I discuss this condition in Chapter 11 on rebooting the endocrine system, but I want to mention it here as well because diet plays such an important role in treatment.

Carbohydrates, especially sugars, stimulate the pancreas to make insulin to regulate blood sugar. Patients with hypoglycemia make an excessive amount of insulin when they ingest carbohydrates, which lowers their blood sugar by significant degrees. For many of my patients, especially those who experience fatigue, brain fog, palpitations, and anxiety when they have not eaten recently, I consider hypoglycemia as another component of their illness. It can easily be diagnosed with a four- to five-hour glucose tolerance test (not done often in conventional medicine today). Hypoglycemia is relatively easy to treat, and treatment often results in some degree of immediate improvement.

Treating or Rebooting a Gastrointestinal System That Is Not Functioning Properly

Rebooting Gut Motility

In his YouTube presentation "The Blood-Brain Axis" (which unfortunately is no longer available there), Dr. Datis Kharrazian does a wonderful job of describing how inflammation in the gut, in the form of inflammatory cytokines, passes directly from the intestines to the brain through the blood-brain barrier (which is not a barrier to cytokines, unfortunately). Once in the brain, sensitive tissues, most notably the vagus nerve and autonomic nervous system, are affected to the point that they cannot function properly. This directly affects intestinal motility because both the vagus nerve and the autonomic nervous system play a major role in regulating peristalsis. With a slowdown in motility, the inflammatory cytokines made in the intestines have even more time to be absorbed into the body and to pass through the blood-brain barrier, which creates a vicious cycle of continuing inflammation and worsening intestinal function. Dr. Kharrazian recommends a fascinating program that can include singing, gargling, and gagging exercises that stimulate the vagus nerve to fire more normally, or "reboot." As unusual as this program sounds, many of my patients have responded well to it. The vagus nerve controls the motility of the upper intestinal tract all the way to the midsection of the large intestine. At that point, other nerves take over. Dr. Kharrazian describes success in using enemas to slowly improve large intestinal and rectal contractility. Please see the section on Polyvagal Theory in Chapter 9 for additional information about rebooting the vagus nerve.

I also have seen excellent results from the use of FSM and osteopathic cranial manipulation (also discussed in Chapter 9) to improve vagal nerve function. Visceral manipulation, which can improve sphincter function, can be helpful as well. The supplement Iberogast, which contains several herbal components, has been of particular benefit for improving intestinal motility in some patients.

Mold toxicity and *Bartonella* have specific effects on motility and other gastrointestinal symptoms, so treating those conditions, if present, is of paramount importance in healing this system. Work on motility (or any other aspect of imbalance) will not "hold" if the root cause has not been addressed.

Rebooting Digestion and the Gut Microbiome

Chronic illness universally impacts the gut microbiome. Given the pivotal role that the gut plays in immune system regulation and hormonal balance, it should be obvious that a person could not be chronically ill with an inflammatory process without the digestive system being affected. I have already emphasized that inflammation weakens the tight junctions of the intestinal cells, regulated by zonulin, predisposing patients to leaky gut and food allergies. Some healthcare providers place an emphasis on looking for leaky gut and then treat it as a central part of healing. While this must be a *part* of treatment, I think it is missing the point. Leaky gut is present in virtually every patient I see, so I am not sure we need to measure or document it. It is not an end in itself; rather, it is an expression of a much deeper imbalance that must be diagnosed and treated. I will again emphasize that looking for and treating the cause is essential. Merely treating the leaky gut itself, though potentially helpful, will not address the core issues.

Inflammation, infection, and toxins in the gut profoundly disturb the delicate microbiome balance, and imbalances are present to some extent in virtually every chronically ill patient. For patients with stronger constitutions, the imbalance may be minor, while for others it is a prominent part of the clinical picture.

> " Leaky gut is present in virtually every patient I see, so I am not sure we need to measure or document it. It is not an end in itself; rather, it is an expression of a much deeper imbalance that must be diagnosed and treated. "

The simplest method of measuring the balance in the gut microbiome is to perform a stool analysis, find out what is deficient (not enough "good" bacteria) or pathogenic (toxic bacteria, *Candida,* or parasites), and treat it. Probiotics can be of great value here. A lot has been written about the use of probiotics, much of it confusing. As mentioned earlier, the main "good" or nourishing bacteria of the small intestines are the *Lactobacillus* species, and the main nourishing bacteria for the large intestines are the *Bifidobacteria* species. While this is an oversimplification, it is a place to start. Both of these species are *anaerobes,* meaning that they can grow only in the absence of oxygen.

For many years, physicians argued about the ideal delivery system for probiotics. Science and opinion seemed to favor the use of enteric-coated capsules, which could move through stomach acid without being damaged and released their contents only when they reached the small intestine, where they could do the most good. It also appeared that refrigerated forms of probiotics, where the bacteria were still alive inside the capsule, rather than "freeze-dried" forms, which could be stored at room temperature, were far superior. This is what I recommended in my previous books.

More recently, however, it has been pointed out that these capsules do not keep air out completely, so, more often than we had realized, the anaerobic bacterial strains end up being exposed to oxygen and are no longer active or alive to the extent we had hoped. While it appears that these dead probiotics have clear health benefits (described in many scientific reports), it would be nice if there were a superior delivery system that would work even better. It turns out that there is. Many bacterial species can *enspore,* meaning that they can surround themselves in a hard casing to protect themselves until they have found a safe home. *Lactobacillus* and *Bifidobacteria* species can now be obtained in spore form so that they are not vulnerable to oxygen and can be released into the small intestine alive and ready to repopulate the biome. I believe that the delivery of these beneficial bacteria in spore form will greatly improve our ability to heal the gut.

Based on the results of stool testing, we also can deliver prebiotics such as fructooligosaccharides (FOS), pectins, and beta glucans, which are nutritional materials that feed the beneficial bacteria and help them grow. These need to be selected carefully, because some prebiotics also foster the growth of toxic bacteria. For example, *Klebsiella,* a common gastrointestinal pathogen, is fed by FOS.

It is beyond the scope of this book to discuss the complicated options for treating intestinal pathogens, but the important point here is that they can be successfully eradicated with a wide array of herbal supplements and occasionally with antibiotics. Antibiotics are a double-edged sword in that they can kill pathogens but also can profoundly affect the delicate microbial balance in the intestines, so they must be used with great care.

The indiscriminate use of antibiotics for many years has created a lot of the problems that we are seeing with the gut microbiome. Even though treating Lyme disease with antibiotics is necessary for healing, it carries the potential for disrupting the biome as well. Most of the time, we can avert problems with the concurrent use of probiotics, but sometimes we unintentionally contribute to a messed-up gut that may be difficult to heal. For a few unfortunate individuals, even going off antibiotics for several weeks or months is not sufficient for the gut to heal. However, in my experience, antibiotics are a critical component of treatment for most patients with Lyme disease or its co-infections, and very few of my patients have had difficulty with their long-term use.

Rebooting with Fecal Transplants

In recent years, we have seen a great deal of interest in the prospect of fecal transplantation. This term describes the process of repopulating a sorely depleted gut microbiome by providing a healthy bacterial population in the form of fecal matter, or stool, via enema, colonoscopy, sigmoidoscopy, or endoscopy.

The difficult-to-treat colitis that is caused by the bacterium *Clostridia difficile* has responded especially well to fecal transplantation, which is a well-accepted form of treatment for this condition. The FDA, concerned that fecal transplantation could be used in inappropriate and potentially dangerous ways, has mandated that the only accepted medical use for fecal transplantation is for the treatment of *C. difficile*. Given that many patients have a seriously disturbed gut microbiome and have not responded well to other forms of treatment, a number of clinics are providing fecal transplants in an effort to help.

I have seen mixed results from the use of fecal transplants. A few patients have been helped, many have noted no particular benefit, and an equal number have reported a worsening of their condition. While fecal transplantation is an exciting concept, I believe we need to learn a great deal more about this treatment process before we can know with certainty who will benefit from it.

The most obvious drawback is to somehow ascertain whose stool is healthy so that it can be used in a transplant. What exactly constitutes a healthy stool, and are we able to measure all of the possible pathogens a stool might contain? Think of this as being similar to a blood transfusion; the donor must be healthy for the blood to be safe. Our ability to determine what is "healthy stool" is still limited, so how certain can we be that we have a safe donor?

The presumed presence of mold/fungus and *Candida* that have colonized the gut in a mold-toxic patient has taught us that until we have dramatically reduced these pathogenic populations, healing may not occur. Treatment for these microorganisms, therefore, is an essential part of healing that can result in a normalization of the gut microbiome.

Rebooting Stomach Acidity and Enzymes

The use of betaine hydrochloride to restore proper acidity to the stomach has been beneficial for many of my patients with disturbed gastrointestinal function. A simple exercise to test for a deficiency of stomach acid (regrettably, we have very few tests at our disposal) is to take one or more capsules of betaine hydrochloride prior to eating a meal and see how you feel. If you experience immediate heartburn, this is clearly not a good strategy. If you feel better immediately, however, taking one or more of these capsules prior to each meal may be beneficial for you.

A host of supplements designed to improve intestinal function have had value for selected patients. The stool testing referred to earlier can help us determine which patients may benefit most from the use of supplemental enzymes. Elevated levels of fecal fat in the stool suggest insufficient lipase, and poor breakdown of proteins suggests that proteases may be helpful. Again, this is such a huge topic that it is beyond the scope of this book. The take-home message is that there are many approaches to rebooting the gut that can be obtained from knowledgeable healthcare practitioners.

Rebooting the CDR and the Microbiome

An exciting observation made by Dr. Robert Naviaux in his landmark 2013 paper "Reversal of Autism-Like Behaviors in Adult Mice with Single-Dose Antipurinergic Therapy" is that when autistic mice were treated with the purinergic blocker suramin, not only did their behaviors improve dramatically, but their gut microbiomes also improved *without any other treatment.* This suggests that when we turn off the cell danger response by curing what is causing it, the biome is capable of righting itself. Clearly this is an intriguing observation that must be explored in greater depth.

Summary

The integrity of the gastrointestinal tract is almost always compromised following chronic illness. This means we need to evaluate gut motility, dysbiosis, food allergies, the stomach's ability to make the appropriate amount of acid, the body's ability to make appropriate amounts of digestive enzymes, and the body's ability to absorb nutrients correctly. Getting the intestines to work well again is an integral part of treatment, but this outcome may be difficult to achieve until the primary causes of illness have been adequately addressed.

CHAPTER 13

Reversing Weight Gain
Bigger Isn't Necessarily Better

The extraordinary proliferation of books devoted to diet and weight loss consumed (pun intended) by the American public is a testament to our obsession with this subject. It is not my intention in this chapter to provide the definitive approach to weight loss, but rather to give an overview of some of the newer approaches that have worked for many of my patients and friends.

Food, as a subject, is laden with meaning on every level, and any attempt to simplify it is unlikely to be useful. Like all habits, eating habits, once ingrained, are very difficult to break. It is much easier to learn something new than to "unlearn" something, and eating is no exception. What, how, and when we eat is tied up in childhood experiences, family patterns, our perceptions of our bodies, how we reward (or punish) ourselves, and our need to fill the empty spaces inside us, and many of us have been taught that food is love. Once these habits and beliefs are firmly established, changing them is no simple task, even when our physicians implore us to eat healthier.

The burgeoning epidemic of obesity is well documented. Many patients with chronic inflammatory illnesses are particularly prone to weight gain that they cannot remove. In the discussion of the Biotoxin Pathway in Chapter 3, I talked about leptin metabolism. In the same way in which insulin resistance develops as an effort by the body to override a decreasing response to insulin by insulin receptors, so, too, does the body strive to make more and more leptin to override a decreasing response by leptin receptors, and this is an important component of mold toxicity and Lyme infection. The increased production of leptin is directly responsible for some of this weight gain, and it is very difficult to repair until the toxicity has been fully addressed and metabolism can be rebooted and normalized. I will discuss this in more detail in just a bit.

The growing realization that eating foods containing additives and preservatives, high-fructose corn syrup, added sugars, sugar substitutes, and hydrogenated fats has contributed to this problem has led to improved labeling and a decrease in the consumption of such products. Despite this effort, there is no evidence of any dent being made in reversing the obesity epidemic.

I know that everyone wants to be told by some expert what, exactly, they should eat to achieve optimal health, but years of clinical practice have convinced me of the remarkable biochemical and genetic diversity of human beings. In essence, your body is your laboratory, and you will need to figure out for yourself which foods agree with you and which do not. Some of us do better as vegetarians, some of us do better as carnivores, and most of us fall somewhere in the middle. So be careful not to implement what you read indiscriminately as if it applies to you personally. It may or may not be correct *for you.* If what you are experimenting with is not working, stop and move on to another approach.

That having been said, let's look at some of the approaches that have had the best results for the patients in my practice, most of whom are struggling with environmentally acquired inflammatory illnesses.

Effective Strategies for Weight Loss

The single most effective weight-loss strategy for many people is to adopt a high-protein, low-carbohydrate diet. There is a bewildering array of these diets out there. The original approach developed by Robert Atkins, MD, and refined by his successors is quite straightforward and effective.

For example, on a personal note, after eating an organic, almost vegetarian diet for four years, and despite running 3 to 4 miles a day, playing basketball, and doing tai chi on a regular basis, I had gained 25 pounds. I was baffled. I was fortunate to begin working with C. Norman Shealy, MD, who casually suggested over lunch that I could not tolerate the amount of carbohydrate I was eating and that I try the Atkins Diet. I immediately did so and lost 20 pounds over a six-month period. I have since experimented with many other diets, but the diet that seems to suit my body best is high-protein and low-carb.

Dr. Atkins rarely gets credit for his pioneering understanding of this way of eating, but the South Beach Diet, the Schwarzbein Principle diet, the ketogenic diet, and the low-amylose diet are similar approaches. Over the years, this type of diet has been the most beneficial for the greatest number of my patients who are working at losing weight.

It is increasingly being recognized that when we eat carbohydrates, our bodies make additional insulin, which has inflammatory effects. (See Chapter 11 for more on the relationship between excess insulin and inflammation.) As we become increasingly insulin resistant—a precursor to diabetes—our bodies make more and more insulin in an attempt to override that resistance and thereby increase the inflammatory process (which I noted previously).

Closely connected to this secretion of excess insulin from the consumption of carbohydrates is that when we eat or drink anything sweet, receptors on our tongues and in our intestines respond to the presence of sweetness by stimulating the pancreas to make more insulin. It does so in anticipation that it is going to have to process what it assumes is sugar or carbohydrates coming down the alimentary canal. So, even though artificial sweeteners and noncaloric natural sweeteners like xylitol and stevia have no nutritional value, they still stimulate the release of insulin in a way that can be detrimental to health and encourage inflammation and subsequent weight gain—an issue that has gained increasing awareness.

More recently, our improving understanding of biochemistry has brought to light several new diets that have an excellent track record so far.

In his book *The Fat Resistance Diet,* Leo Galland, MD, focuses on how to lower leptin levels in the body by reducing the inflammation that triggers the production of leptins. This issue is of particular relevance to patients with mold toxicity. One of the central features of the Biotoxin Pathway is that mold toxins specifically interfere with the metabolism of leptins, creating leptin resistance, in which the body makes more and more leptins (from adipocytes, or fat cells) in an effort to override that resistance. Leptins are important molecules that signal to us that we are full after eating. If our bodies cannot respond to that signal properly and we do not have the sensation of satiety, we keep eating. Dr. Galland details which foods should be avoided to prevent the promotion of an inflammatory response and which foods should be featured to encourage an anti-inflammatory response (for example, foods that are high in omega-3 fatty acids and fiber).

> **" It is increasingly being recognized that when we eat carbohydrates, our bodies make additional insulin, which has inflammatory effects. "**

Steven R. Gundry, MD, outlines a different approach in his book *The Plant Paradox.* Dr. Gundry focuses on lectins (whose name is confusingly similar to leptins), which essentially are toxins made by plants for the purpose of dissuading animals from eating them. Over the millennia, plant chemistry has evolved to include the ability to make lectins, which are unpleasant-tasting or even toxic to the animals that consume them. Our diets are filled with lectins, including such foods as legumes, grains, and certain vegetables (especially those with seeds in them, like tomatoes and cucumbers), and Dr. Gundry believes that, unbeknownst to us, our consumption of these lectins has been making us sick. Following a low-lectin diet can minimize those toxic effects, and he has demonstrated clear benefits of this type of diet in losing weight, controlling diabetes, and reversing autoimmune disease.

A different approach can be found in Dr. Michael Mosley's book *The FastDiet.* He reviews the scientific evidence for fasting as a method of reversing inflammation (almost everyone keeps coming back to inflammation as the source of our difficulties), losing weight, and improving health, especially when combined with high-intensity interval training. His method is fairly easy to do: two days a week, you simply do not eat for sixteen hours at a time, and during the remaining eight hours of the day you eat limited amounts of food—600 calories for men and 500 calories for women. I have found this to be a practical and successful approach to weight loss for many of my patients (and myself).

Let me restate my suspicion that our growing problems with obesity are directly connected to the increasing levels and types of toxins to which we are exposed. The vast proliferation of chemicals in our environment (most of which have not been tested for their effects in humans), coupled with the increased use of chemical additives in foods and our exposures to heavy metal toxins and all kinds of electromagnetic waves never before present in human existence, are unavoidable. We are all at risk. There is no remote island to which we can escape. Our bodies must process these substances and energies, which distract them from more basic functions, and many especially sensitive people are predisposed to illness, including weight gain, as a result. Unless we do something about our environment, we will reach the point of no return before we know it.

I suspect that obesity is only the tip of that iceberg, and that the illnesses discussed in this book are a part of that iceberg. Remember the *Titanic.*

> " **Our growing problems with obesity are directly connected to the increasing levels and types of toxins to which we are exposed.** "

CHAPTER 14

Rebooting Methylation

With an Emphasis on Unusually Sensitive Patients

While methylation is an important component of health and a common problem for patients sufferng from chronic illnesses, it has been receiving an unusual amount of attention lately, elevating its importance to heights that are, perhaps, out of proportion to the role that it plays. I am devoting a chapter to methylation in the hope that I can help put the whole concept into perspective.

What Is Methylation?

While it may sound complicated, methylation is simply the biochemical process of adding a methyl group to a molecule. A methyl group consists of a carbon atom surrounded on three sides by hydrogen atoms, allowing the fourth side (to the left of the central carbon atom in Figure 14.1 below) to bind to another substance.

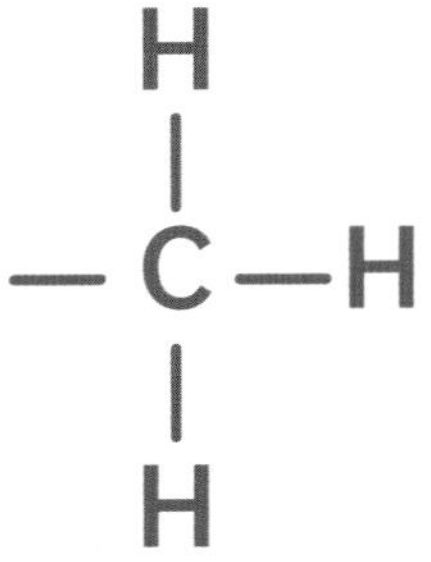

$$
\begin{array}{c}
H \\
| \\
-\,C\,-H \\
| \\
H
\end{array}
$$

Figure 14.1. A methyl group.

The process of methylation is basic to hundreds of important biochemical reactions in the body. The most important of these is creating glutathione, which is a central material in creating energy, detoxifying, and dealing with free radical buildup. The body uses methylation itself for detoxifying, repairing damaged DNA, converting serotonin to melatonin, and many other critical processes.

The most common symptoms of methylation deficiency, or an inability to methylate properly, are fatigue, cognitive impairment, pain, and insomnia.

Let's begin with a little history of how methylation rose to the top of the to-do lists of functional medicine physicians.

How Methylation Came to Prominence

Rich Van Konynenburg, PhD, spent his career as a pioneering engineer who became a major troubleshooter for the Livermore National Laboratory. After his retirement, he became fascinated by the difficulties faced by patients with myalgic encephalopathy/chronic fatigue syndrome (ME/CFS) and fibromyalgia. He spent several years researching the biochemistry of these illnesses, and in 2003 he wrote a seminal paper suggesting that methylation chemistry, when not functioning properly, could explain virtually all of the clinical findings in ME/CFS. This hypothesis was developed from the work of Amy Yasko, PhD, who was working with autistic children. Dr. Yasko suggested that autism was the form that illness took in younger patients and was similar to ME/CFS and fibromyalgia in adults and to neurodegenerative diseases in older patients.

Dr. Van Konynenburg was still having trouble convincing physicians to take his work seriously when I heard him speak at a medical meeting in 2007. I was so intrigued by the logic and simplicity of his ideas that when I got home from the meeting, I immediately placed fifty-one patients with ME/CFS and fibromyalgia on the five supplements that Dr. Van Konynenburg had developed from Dr. Yasko's work and that he described as a "simplified methylation protocol." Essentially, this protocol consisted of hydroxocobalamin (hydroxy-B12), a tiny dose of 5-methyl tetrahydrofolate (5-MTHF), and several other supplements to support methylation chemistry.

After several months of taking these supplements, 70 percent of my patients noted definite improvements, and 20 percent reported that they felt *markedly* better. This was the first time that Dr. Van

Konynenburg's hypothesis had been tested, and he was delighted with our initial results.

Dr. Van Konynenburg and I were fortunate to receive private funding for a more elaborate clinical study, and we entered thirty patients, all of whom met the criteria for ME/CFS and fibromyalgia, into a research project in which we measured their methylation chemistry and kept a detailed clinical symptom diary before giving them supplements. We gave all of them exactly the same supplements I had used in the pilot project and followed their clinical progress closely for nine months, measuring their methylation chemistry with a blood test from Health Diagnostic Laboratory every three months. The results were impressive. Despite the unusually long trial period, almost all of the patients participated for the entire duration. Not a single patient who began the study had normal methylation chemistry, suggesting that Dr. Van Konynenburg's hypothesis about abnormal methylation chemistry playing a significant role in ME/CFS and fibromyalgia was valid.

By the end of six months, glutathione and SAM had risen to normal levels in almost every patient. After nine months, they were better still. More important, the patients had improved clinically: 77 percent reported improved energy, 65 percent reported improved sleep, 73 percent reported improved mental clarity, and 54 percent reported a decrease in pain. To put this differently, 83 percent told me that they had improved (defined by them as being 15 to 50 percent better), and of those who had improved, 27 percent reported that they were *much* better, defined as 50 to 100 percent better. The average improvement at six months was 48 percent and was statistically significant.

Dr. Van Konynenburg and I felt that we had begun to address the important questions: *Does methylation play a role in ME/CFS and fibromyalgia,* and if so, *Is it treatable?* The answer to both questions was a resounding yes!

The funding for this project did not, by design, include an institutional review board, which would have allowed us to submit these findings to a peer-reviewed journal. You can find the details in my book *Healing Is Possible* and in the December 2011 issue of the *Townsend Letter.* Dr. Van Konynenburg and I presented this work at quite a few medical meetings, both national and international, and it received a good deal of attention. What followed was an explosion of interest in methylation as an important component of health and illness, and it came to be viewed as a central area that needed to be addressed clinically. After all, the end product of methylation, as I have emphasized, is glutathione, which is essential for energy, detoxification, and oxidative chemistry. So methylation appears to be something that needs to be addressed early on in treatment, perhaps for everyone who is chronically ill.

Well, yes and no. As you might expect, in complex patients, it is more complicated than that.

Timing Methylation Treatment Correctly

In Chapter 8, I discussed the cell danger response (CDR). One important component of the CDR is that when a cell feels threatened, it intentionally shuts down methylation to deprive an infecting microbe of the capability to hijack the body's own methylation chemistry so that the microbe can reproduce. It therefore should come as no surprise that virtually all chronically ill and inflamed patients, in whom the CDR has been triggered, have measurably low methylation chemistry. But does that mean we should immediately provide all patients with the supplements to maximize methylation?

It turns out that the answer to that question is: *It's all about timing.*

Because the CDR is a protective mechanism, overriding that mechanism before the body is ready to move forward might not be a great idea, especially for sensitive patients. In my experience, approximately half of sensitive patients, if given vitamin B12 or folate (usually in the form of 5-MTHF), experience an intense worsening of their symptoms—even when given minuscule doses. I would like to emphasize that in our study of thirty patients, *supplementing with only 200 micrograms of 5-MTHF improved every patient's ability to methylate.* Some practitioners are recommending massive dosages of folate with the idea that if some is good, more is better. This is rarely true. In my practice, in fact, the opposite is far more common. Many of the patients who come to see me are on dosages of folate that are way too high, and they improve when we reduce the dosage dramatically or stop the supplement altogether.

Unfortunately, some practitioners insist that because methylation is so important, especially to detoxification, patients must continue taking the supplements. Worse, they advise increasing the dosage even when a patient is clearly reacting badly. This is not wise. The vast majority of these patients will get worse—often much worse. The body is saying that it is not yet ready to begin this area of treatment. I have found, however, that once a patient has eased into treatment of the underlying main issue (usually mold toxicity or Lyme disease), he or she can take methylation supplements comfortably and with benefit later in the course of treatment.

I worry that healthcare providers have escalated the importance of methylation in detoxification to the point that they are telling patients that if they can't methylate, they can't get better, so they have to keep taking their methylation supplements no matter how bad they feel. Not so! Please keep in mind that the body has many avenues and organs of detoxification, so if methylation can't be started early on, the body can and does utilize other systems to detoxify. Sensitive patients who cannot take methylation supplements yet can begin treatment, improve, and *then* initiate efforts to methylate better.

Another area I would like to address is the love affair that integrative practitioners are having with glutathione supplementation. While there is no question that glutathione is an essential component of health, its indiscriminate use is not always justified.

Many patients with mold toxicity, for example, get worse when given glutathione in any form—oral, IV, or topical. You may recall that in Chapter 3, I encouraged the use of oral glutathione as a prevocational agent to make urine testing for mycotoxins more accurate. At the same time, I cautioned that in sensitive patients, glutathione mobilizes molds (and other toxins) faster than their bodies can process those toxins, and the patients become more toxic. They experience this increase in toxicity as an intense exacerbation of their usual symptoms. This is not a reaction that these patients can override by pushing themselves to "ride it out," although many of them try anyway. It is important to recognize when this adverse reaction is occurring and to stop taking glutathione until the body is able to right itself. Many of my sensitive patients simply cannot take glutathione in any form until later in treatment when they are on the mend.

A second concern I have is that many practitioners do not seem to realize that glutathione, as important as it is, is carefully monitored and regulated by the body. When glutathione is taken in any form, the body's biofeedback systems announce that it has plenty of glutathione on board now, so it can stop making glutathione using the process of methylation. What this means is that many patients who are taking glutathione are unwittingly *turning off their ability to methylate,* which is counterproductive in patients who already are not methylating well.

This is not to say that glutathione should not be used in treatment, but it is a plea to practitioners to make an effort to discern which patients might benefit from the use of glutathione and which should

> **" I have found that once a patient has eased into treatment of the underlying main issue (usually mold toxicity or Lyme disease), he or she can take methylation supplements comfortably and with benefit later in the course of treatment. "**

not be offered glutathione until they have improved. I have intentionally emphasized difficulties with methylation and the use of glutathione because these difficulties are common in my sensitive patients. I also want to make it clear that the use of glutathione in patients who have a stronger constitution can be quite helpful. Janette Hope, MD, has described the use of intranasal glutathione being of great help in treating patients with mold toxicity. Many patients have found immediate and clear benefits from the use of oral or intravenous glutathione as well. My take-home message is simply to alert practitioners and patients to the possible downside of the use of glutathione and to advise you to be very careful when instituting treatment with this material.

The bottom line is that sensitive and toxic patients must approach methylation very carefully, and they and their healthcare practitioners must be mindful that this area of treatment may need to be postponed until they have begun to improve. While I believe that methylation is indeed important (heck, I helped put it on the map!), I urge patients and physicians to put its value into perspective and to always look for the greater context in which we view treatment.

CHAPTER 15

Rebooting the Body's Ability to Detoxify

Cleanliness Is Next to Godliness

Throughout this book, I have emphasized the importance of detoxification in the healing process. Early on, in Chapter 2, I talked about the need to specify, as much as possible, exactly which toxins are present. Many publications describe detoxification as a key issue, but they seldom address the fact that the body metabolizes different toxins in different ways, using different pathways, leaving readers with the idea that every form of detoxification is globally useful. Not so. In Chapter 3, I tried to be as detailed as possible about which detoxification strategies work best for which mold toxins, pulling together the best information currently available.

This chapter is not intended to be a comprehensive review of this subject, but to remind you of the importance of detoxification and to give you a CliffsNotes–style survey with an emphasis on the procedures and techniques that I have found most beneficial for my toxic and sensitive patients. There are many additional approaches to detoxification, but I have not seen those approaches deliver specific value for my patients; I am not aware that they have been specifically applied to the major toxicities and infections discussed in this book; and, when applied indiscriminately, they often have made my patients worse. So, if I do not mention your favorite process here, it does not mean that it will not work for you. I am simply sharing what has worked best for my patients.

Detoxification Overview

We can think about detoxification from an organ perspective or from a biochemical perspective. First, let's take the organ perspective. The organs and systems that remove toxins from the body include the liver, intestinal system, kidneys, lymphatic system, skin, and lungs (see Figure 15.1).

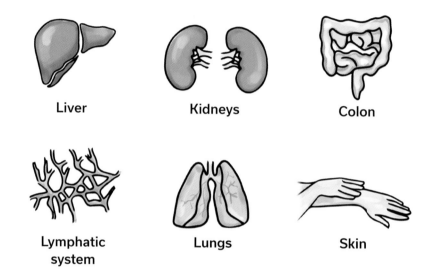

Figure 15.1. These are the organs and systems that remove toxins from the body.

Liver   Kidneys   Colon

Lymphatic system   Lungs   Skin

The biochemical perspective refers to the precise mechanisms by which a toxic material is processed by the body, primarily the liver. As I have emphasized, toxins often interfere with the ability of these organ systems to function, paradoxically creating a downward spiral—the more toxic a person becomes, the less able these organs are to get rid of toxins. Therefore, detoxification needs to be considered early on in treatment, or these difficulties might hijack our efforts. There are some basic strategies that are common to many approaches described in detail in many other books, but I will summarize them below. From a biochemical perspective, we have learned a good deal about what are described as the Phase I, Phase II, and Phase III detoxification pathways in the liver, and there are some basic strategies common to many treatment approaches that I will review briefly as well.

One of my biggest concerns about all forms of treatment is that toxic and sensitive patients are easily overwhelmed by all the things they think they *must* do to get well. Once overwhelmed, they often feel like giving up because they do not see how they can possibly do everything being asked of them. They often get the idea that if they don't do it all perfectly, they cannot get well. This is simply not true. A steady, consistent application of even a few of these strategies will, over time, go a long way. That is why I approach treatment very slowly and am careful to assess each patient individually, trying to ascertain his or her personal limits. While I can easily point out dozens of methods that might be helpful, I try to make each suggestion as specific to that particular patient's abilities as possible. For example, patients who are constitutionally stronger may not need some of these methods because their bodies are already keeping up with the toxic load. For the most compromised patients, we must proceed very slowly and cautiously.

Detoxification Basics

Food

As discussed in previous chapters (especially Chapters 12 and 13), each person needs to experiment with his or her diet to find what works best. Start by figuring out if any foods are directly harmful, which includes looking at allergies and at mast cell activation. People experience a wide range of reactivity to foods, from severe anaphylactic reactions to peanuts or shellfish to minor issues such as mild water weight gain. Gluten and dairy sensitivities are extremely common in my patients.

Once it is clear that a particular food triggers a reaction, it may not be necessary to distinguish true allergy from reactivity, since in either case that food should be avoided as much as possible. However, the luxury of "as much as possible" does not apply to a true allergy or, in the case of gluten, full-blown celiac disease, in which even a minute amount of gluten can cause a severe reaction.

For a few patients, a low-histamine or low-oxalate diet may be helpful. You can find an overview of low-histamine diets on page 110. A small percentage of patients are unusually sensitive to foods that contain oxalates (such as kale or spinach), and this sensitivity will contribute to the global inflammatory process that is so central to our patients' woes.

Other food-related considerations come into play as well. Eating organically grown and raised foods (which includes avoiding genetically modified, or GMO, foods) can make a huge difference to sensitive patients. Admittedly, buying organic ingredients can be expensive. It can also be critical to healing. Other basic strategies that are often stressed as being helpful include chewing food slowly and thoroughly, avoiding overeating, and making mealtimes relaxing events (meaning, please don't listen to the news or use electronic gadgets while eating!).

Hydration

The kidneys require adequate fluid to flush toxins, and I urge those who are not moved to consume much liquid to consider this point carefully. That being said, while many authorities insist that everyone should drink eight 8-ounce glasses of water per day, this recommendation must be individualized. This quantity of water would make some people feel bloated and uncomfortable. There is a medical condition called *water toxicity* in which too much water can provoke serious disturbances, so each patient must tailor his or her liquid consumption to his or her own comfort level.

Try to avoid drinking water from plastic containers; many plastic components are xenoestrogens, which compete with estrogen in the body and interfere with many aspects of metabolism. If possible, drink your beverages from glass or ceramic containers instead.

Some patients may need to add electrolytes to their water to optimize balance, especially if they are using a sauna or Epsom salt baths on a regular basis, which can deplete the body of sodium and potassium, along with other minerals.

Sweating

Sweating is an important part of detoxification. Unfortunately, some toxic patients have autonomic nervous system dysfunction that interferes with their ability to sweat. Many of my patients initially report that they are unable to perspire and notice that as their health improves, their ability to sweat improves as well. Sensitive patients may not be able to tolerate heat for very long, so each person needs to determine how much he or she can do that does not worsen his or her condition.

Saunas, hot baths, hot tubs, sitting in the sun—anything that encourages sweating may be helpful in moderation. Low-temperature infrared saunas have been particularly helpful for many patients, but their use should be limited to two or three times a week. Overdoing the sauna can lead to increased toxicity and the depletion of electrolytes and minerals, as noted earlier. For especially sensitive patients, foot baths with Epsom salts can be useful as well.

Exercise

Movement of any kind gets fluids circulating and is of great benefit. Because many patients, as part of their illness, have post-exertional malaise or myalgia (pain), exercise must be prescribed carefully. Any kind of post-exertional issue means that patients are *using up energy reserves they don't have*. This is not a viable strategy. Well-meaning friends, relatives, and even physicians often say, "If you would only exercise, you would get well." This is terrible advice. If you keep using up energy reserves you don't have, you will become more and more depleted and cannot recover. Only when you have recovered to the point that you have reserves can you begin to exercise in earnest. While sick, it is important to understand that you can exercise, but only to the point that you do *not* get worse afterward. The amount of exercise you can do may be minuscule (for example, walking to the mailbox and back), but as long as you incorporate some gentle movement without overdoing it, you can slowly improve.

Sleep

Restorative sleep is very important to the healing process. Unfortunately, most of my patients have a lot of difficulty both falling asleep and staying asleep. Even when they spend twelve hours in bed, they often awake unrefreshed (a cardinal symptom of fibromyalgia). Keeping to a regular sleep schedule, minimizing light in the bedroom, avoiding naps as much as possible, and avoiding caffeine and alcohol before bedtime are basic strategies to improve sleep. Supplements and medications may be needed to improve sleep to facilitate healing.

Many of my patients have unrecognized sleep apnea, a condition that involves episodes during sleep in which a person stops breathing until the increase in carbon dioxide created by a lack of ventilation

"While sick, it is important to understand that you can exercise, but only to the point that you do not get worse afterward."

forces them to breathe again. Going without oxygen for variable periods of time is antithetical to health and healing. Any patient who snores (your significant other will tell you if you are one of them!) should be evaluated with a sleep study. Because sleep apnea can be treated with the use of a CPAP machine or a dental appliance, thus improving oxygen supply all night long, sleep can be an important area to evaluate early on in treatment.

Detoxification by Organ System

Now let's look separately at the major organs that provide us with the ability to detoxify: the liver, gallbladder, gastrointestinal system, kidneys, lymphatic system, and skin.

Liver

The liver is the body's major organ of detoxification. Especially for many of the toxins discussed in this book, it is *the* key player. Basic strategies of eating organic foods, steering clear of GMOs, and avoiding exposure to plastics, chemicals, and other toxins are important, since that minimizes the extra work the liver must do when harmful substances are on board. For a variety of reasons, avoiding sugar substitutes falls into this same category. Having to process Splenda and/or NutraSweet is a burden on the liver, and both of those sweeteners stimulate the pancreas to make additional insulin, which adds to the common complication of insulin resistance. (See page 207 for more on insulin resistance.) The cleaner your diet, the more energy your liver has to get rid of toxins.

Several supplements have proven to be of particular benefit for the liver. Milk thistle (one capsule two or three times daily) and alpha-lipoic acid, or ALA (200 milligrams twice daily) are two of the most useful.

A variety of tinctures and combinations have worked well for my sensitive patients, who may need to start with a single drop and slowly increase the dosage as tolerated. My two favorites, both of which are taken twice daily, are Tox-Ease, either the capsules or the GL (a liquid formulation for very sensitive individuals), and apo-Hepat drops, which are a little stronger.

Gallbladder

For many years, it has been thought that the main function of the gallbladder is to concentrate bile to make it more efficient in the digestion of fat. Recently, however, a newer, more complex appreciation of the importance of bile has grown steadily. Bile is a primarily aqueous secretion (it is 95 percent water) that originates in the liver cells (hepatocytes) and is *modified* by absorptive and secretory systems in the bile duct epithelial lining. We now realize that bile is the major route of excretion for toxins (especially those that are soluble in lipids, or fats) and helps protect the gastrointestinal tract by secreting immunoglobulin A and stimulating the innate immune system in the intestines. When these secretions are impaired (which is certainly relevant for those who have had their gallbladders surgically removed), as they are in the presence of many chronic inflammatory illnesses, the body's ability to detoxify is significantly hindered, even when the correct binders are provided.

Efforts to improve the formation and secretion of bile, therefore, may be essential to the detoxification process. I encourage you to listen to Scott Forsgren's podcast with Kelly Halderman, MD (found at www.betterhealthguy.com/episode79), in which Dr. Halderman does a brilliant job of describing this process and suggesting supplements to help the body make and secrete bile more effectively. These include milk thistle, pantothenic acid (vitamin B5), acetyl-L-carnitine, bitters, phosphatidylcholine, and ox bile. I worked for many years with Wayne Anderson, ND, who often added the medication ursodiol (Actigall) to improve bile flow in his patients.

Returning briefly to our understanding of the cell danger response, Dr. Naviaux reminds us that hepatocytes express purigenic receptors on their plasma membranes, meaning that bile secretion is partly regulated by the effect of nucleotides and nucleosides on these membranes. In his published research on autistic mice, he found decreased bile acid secretion that normalized when the mice were treated with suramin. While this information is highly technical for some readers, what it boils down to is that *bile secretion is really important*. When inflammation is not controlled, the body's ability to make bile is compromised, and when inflammation is healed, the body's ability to make bile improves.

I believe that integrating our new understanding of the importance of the secretion of bile into the treatment of toxicity will provide additional improvement for patients who are having difficulty with detoxification.

Gastrointestinal System

Because I have devoted a separate chapter to the gastrointestinal system, I won't review that information here, but this system needs to be included in our discussion of detoxification. Constipation is a particular problem for my patients because toxins that remain in the intestines longer than they should can be reabsorbed, which is clearly counterproductive to treatment. Therefore, constipation needs to be addressed early on in treatment. The toxins themselves stimulate the production of inflammatory cytokines, which in turn trigger inflammation in the central nervous system, especially affecting the vagus nerve and autonomic nervous system, which interferes with intestinal motility and causes constipation. Exercises to improve vagal nerve function, osteopathic cranial manipulation, and FSM can be helpful here. I have had a number of patients who had never fully recovered from a surgical procedure. When the appropriate FSM frequencies were used to "remove" anesthesia toxicity, several of those patients reported remarkable and rapid recoveries.

Kidneys

As described earlier, drinking adequate amounts of liquid is essential to flush out certain toxins. Some compromised patients benefit from intravenous preparations. My favorite supplement to assist the kidneys in functioning better is the homeopathic remedy Renelix, five to ten drops twice daily (starting with one drop for sensitive patients). The use of FSM frequencies for the liver, intestines, kidneys, skin, and lymphatic system also has been helpful for many patients. Some practitioners emphasize the need to alkalinize the body by using an alkaline diet or alkaline water to promote detoxification, but I have not found this approach to be useful or necessary.

Lymphatic System

The lymphatic system (see Figure 15.2) is a complete circulatory system that helps drain waste products and infectious agents from the body. The lymph channels drain into the lymph glands and ultimately into the superior vena cava on the left side of the chest, connecting with the circulatory system, which clears these toxins through the kidneys and liver.

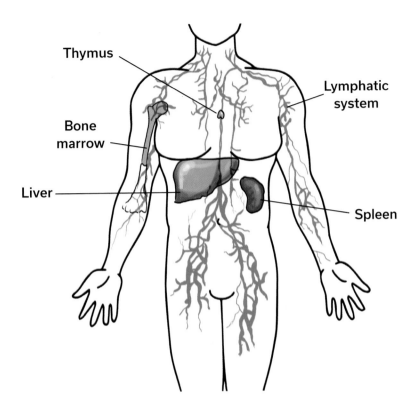

Figure 15.2. The lymphatic system helps drain waste products and infectious agents from the body.

An unusually high percentage of toxic and sensitive patients have sluggish or blocked lymphatic drainage. When flow is restored, many of them begin to improve. Referring these patients to therapists who are trained in lymphatic massage can be very helpful. I also have seen excellent responses to the homeopathic remedy Itires at a dosage of five to ten drops twice daily (starting with just one drop for sensitive patients). Again, FSM using the lymphatic and lymph gland frequencies has been helpful. A simple exercise of flexing the feet and hands ten times, several times a day, helps circulate lymph, and movement of any kind is beneficial for improving lymph circulation.

Skin

Sweating, as described on page 234, is really helpful for releasing toxins. Dry brushing also is often beneficial in this regard. Applying a soft but firm brush over large areas of the skin once or twice a day helps remove dead skin cells and improves lymphatic flow.

As alluded to in Chapter 3, one of my favorite treatments for mold toxicity is the use of intravenous phosphatidylcholine. In addition to being gentle and effective at removing toxins directly, it helps heal the damaged membranes to which those toxins are attached. I am grateful to my friend and colleague Kristine Gedroic, MD, who has a great deal of experience with this treatment, for providing the following discussion.

As director of the Gedroic Medical Institute in Morristown, New Jersey, I see patients with all kinds of complex chronic illnesses. Many years ago, when my patient population expanded to include different neurological disorders—seizures, neuropathies, nerve palsies, and everything nerve-related—I needed answers as to how to better help them. It was at this time that I first heard of Dr. Patricia Kane, a "biochemical master" of treating neurological problems. Dr. Kane's approach involves testing patients with a comprehensive metabolic panel (CMP), complete blood count (CBC), and lipid panel (cholesterol panel), along with a fatty acid analysis through Johns Hopkins' Peroxisomal Disorders DNA Diagnostic Laboratory. Together, this data allows for a customized road map on how to supplement patients in order to regain neurological balance as well as cell membrane stability, which go hand in hand. In severe cases, the Patricia Kane Protocol, or "PK Protocol" as it's commonly called, can be administered intravenously to resolve many neurological problems. I soon learned that this therapy works incredibly well for my mold patients, too.

The question to be answered, then, is, "Why does the PK Protocol work?" Cellular membranes are a mixture of fats—cholesterols, sphingolipids, and phospholipids—along with embedded proteins. When a cell is balanced with respect to its fatty acids (a balance the PK Protocol helps to restore), it is healthy. It is as simple as that. Some scientists have likened the cell membrane to the "keyboard" and the cellular DNA to the "hard drive." All the messages that the cell sends to neighboring cells emanate from the membrane, or keyboard. Therefore, when the cell membrane is unstable or is "missing keys," the messages being sent out are skewed. Mold exposure, quite simply, messes up a lot of keys.

When we test patients for fatty acid balance, the detection of sphingolipid and omega-6 fatty acid deficits often directs us to suspect the presence of mold toxicity. Mold exposure drops levels of these fats by blocking the enzymes needed to make them, and in doing so wreaks havoc on the cell membrane, causing high levels of membrane instability. This is like a circus tent missing a pole, causing the tent (or membrane) to flap around. This is particularly important as it relates to the nervous system because it is a system comprised mainly of fat. The myelin sheath that protects a nerve is literally a long fatty membrane. Therefore, if there are significant fatty acid deficits, the myelin sheaths are affected, too.

The degree of fatty acid deficit often correlates to the severity of the neurological symptoms. These symptoms include, but are not limited to, numbness, tingling, weak hand strength, double vision, pins-and-needles feelings throughout the body, clumsy gait, and bumping into things. I describe cell membrane instability to my patients by explaining that nerves are like copper wires missing their protective sheaths (aka membranes), causing sparks to fly everywhere.

By identifying the kinds of fatty acid deficits a patient is coping with, along with the degree of fatty acid deficiency, targeted supplementation—both oral and, at times, IV—can be given to correct these imbalances. When I began to supplement my mold patients in this way, many finally started to recover.

In treating mold cases, I often begin with one to two months of nothing but supplementation of the missing fats—along with electrolytes, minerals, and amino acids—in order to help their bodies begin to rebuild. Once they are "stabilized" with this approach, further treatment can begin more successfully with much less reactivity.

While many, if not all, patients clearly benefit from oral fatty acid supplementation, many require IV treatment with the PK Protocol to achieve the greatest clinical outcome. This is due to either the severity of the fatty acid deviations—which require IV care to treat the patient quickly—or the fact that epigenetic elements are being corrected by the use of the IV. Since molds produce large amounts of volatile organic compounds, or VOCs, it is known that both mycotoxins (the toxic chemicals produced by mold) and these VOCs—which can trigger chronic inflammation—bind to cellular DNA and produce broad-reaching epigenetic effects or disturbances in DNA regulation. I have found that the use of IV phosphatidylcholine through the PK Protocol has the ability to resolve these epigenetic effects and improve DNA regulation.

Allow me to illustrate with Adam. This young adult patient came to see me after having received ten years of on-and-off antibiotics for "chronic Lyme disease," yet all of his symptoms—extreme anxiety, OCD, perseveration, insomnia, pins-and-needles sensations, hypersensitivity to noise and light, and gastrointestinal distress—were neurological symptoms that seemed to be related to mold exposure. We performed fatty acid testing through the Johns Hopkins Lab, and, sure enough, his fatty acids had all sorts of deviations. Nothing was in balance, which accounts—at the cellular level—for why Adam felt terrible. We began Adam on one month of oral fatty acid supplementation, and within four weeks, he began to feel a little less edgy. I also encouraged him to investigate his home environment since I felt that mold was his primary problem—and not the Lyme disease for which he had received years of antibiotics. It turned out that Adam's apartment didn't have just a little mold; the drywall was completely saturated with mold, and he needed to move. No amount of remediation would have made his living environment safe for him. We immediately began IV treatment with the PK Protocol, and within ten treatments, Adam began to feel better. His anxiety lessened, his insomnia improved, the numbness and tingling decreased, and just about every symptom he had been experiencing was slowly getting better. Adam moved, and then he proceeded with an additional ten IV treatments, after completion of which he noted, "I feel the best I have in over ten years!" He then maintained oral care and slowly tapered off treatment. Today, I still see Adam, but only once or twice a year. He is healthy, happy, and living a good life again. It is wonderful to have cases like his. We have helped hundreds of patients with mold illness recover a normal quality of life.

As providers to the very sick, we need to have as many tools as possible to treat the critically ill and sensitive patients that we see each and every day. I hope that by sharing my experience with this very successful PK therapy, more patients will be able to find providers who will discuss this option for their future care.

In good health,

Kristine Gedroic, MD

Summary

As you can see, there are many, many approaches to detoxification, and I have been able to discuss only some of them in this chapter. The take-home message is that detoxification is an integral part of treatment and must be considered as a part of the overall game plan.

In the next chapter, Bob Miller and his son Matthew will add to this discussion. Bob has studied the genetic components of detoxification and will share his observations about specific imbalances that need to be addressed in compromised patients.

How Your Genetic Makeup May Affect Your Ability to Detoxify

by Bob Miller, CTN, with Matthew R. Miller

Despite the excitement that our new ability to measure genetic information has generated, it is not yet clear to what extent that information can guide diagnosis and treatment. This genetic information is being provided in the form of single nucleotide polymorphisms, or SNPs—pronounced "snips"—and is readily available from a variety of sources; 23andMe and Tree of Life Solutions are two examples. What the public does not fully understand is that these tests measure genetic *potential*, and it is not always clear whether those genes are actually being *expressed*.

As Bob and Matthew Miller will explain in this chapter, genes can be turned off and on by a wide variety of stimuli. If a gene is turned on, we say that it is expressed. Simply *having* a SNP does not mean that the genetic potential is being realized, but many patients are under the impression that having a SNP gives their bodies little or no choice as to how their biochemistry will be affected. This is not the case. This chapter is intended to help patients and physicians better understand how this information can be used to evaluate *possible* biochemical imbalances and influence them with the use of natural materials that can help make these biochemical reactions work more smoothly and override potential genetic influences.

This is the key point: understanding the interactions between SNPs and biochemistry allows us to minimize these imbalances. This can be of particular benefit for the most sensitive and toxic patients. Until I heard Bob Miller deliver several lectures on this subject recently, I was unaware that our understanding had come this far, so I am delighted to be able to offer you the opportunity to learn about this approach.

I confess that I fibbed when I told you the section on the cell danger response would be the last of the biochemistry that this book would offer. It is not possible to grasp the interactions between SNPs and chemistry without having it laid out for you, and in the following pages, Bob and his son Matthew have done a wonderful job of making this information clear.

Spoiler alert: For those of you who have access to your genetic information, you may want to check your SNPs for MTHFR A1298, ME1, HFE, SLC40A1, CP, SOD1, SOD2, SOD3, CAT, GSR, ULK1, ULK2, and ATG13.

Now, Bob and Matthew will lead us, step by step, through an understanding of how to work with our genetic information in a way that may help us "reboot" some of the major biochemical blockages to which our bodies are subject.

Perhaps you have been in an environment where you were exposed to mold, chemicals, cigarette smoke, or perfume. You may have experienced a headache, nausea, or sickness that lasted for days. Or maybe you feel as though no matter what you try, you are unable to detoxify properly. People are often told, "It is all in your head," or even ridiculed for their reactions to these exposures. What's the difference for some individuals? Although many factors are in play here, the genetic pattern that you inherited from your parents may be a significant contributor to your ability to detox.

You may be thinking, "Genetics—isn't that what determines the color of my hair, my height, and my predisposition to disease?" Yes, that's true, but your genetics also determines how you might create extra free radicals that can be harmful, or how you might make less-than-optimal antioxidants to protect yourself from damaging free radicals. Genetic characteristics, along with environmental factors, also might influence a cellular cleansing process called *autophagy* to perform less than optimally when clearing toxins.

This concept is often referred to as *functional genomics:* observing how genetic patterns affect protein and enzymatic function, which may determine how efficiently you detox, how susceptible you may be when exposed to dangerous pathogens and environmental toxins, and ultimately how fast you age.

You Are Not "Defective" or "Dirty," and You Are Not "Doomed"

I have provided health coaching to thousands of people regarding functional nutrition and their genetic patterns, and I often see fear in people's eyes that they are going to learn some very bad news that "dooms" them. I like to tell them that where I am looking, I have no *bad news*. The goal is to locate where you may have inherited some *potential* genetic weakness that could:

- Influence your ability to create antioxidants that protect you from free radicals and toxic substances.
- Increase levels of free radicals that could potentially cause damage.
- Impair your ability to clean up cellular debris via autophagy.

The great news is that this information *empowers* you to compensate by implementing lifestyle changes, dietary changes, and precision nutrition from vitamins, minerals, and botanicals to support this potential weakness. In my opinion, ignorance is not bliss; knowing where your weak spots are and proactively compensating when needed is reassuring. The perfect functional genetic pattern does not exist. But you can take the hand you were dealt and support it with commonsense approaches.

Free Radicals, Antioxidants, Toxicity, and Inflammation

Before we begin looking at genetic patterns that have the potential to impair your ability to detox, let me briefly explain the fundamental concept of traditional naturopathy. It goes back to the scientist Antoine Bechamp (1816–1908), a contemporary of Louis Pasteur (1822–1895). Pasteur developed the germ theory and proclaimed that germs are the cause of illness; the basic idea is that if we eradicate the germs, health will follow. Bechamp said, "Louis, you are right, there is a germ there, but it's the terrain, or internal environment of the body, that creates conditions where the germ can thrive." In other words, if your body is congested, toxic, and inflamed, pathogens find it easier to live inside you.

A good example is shingles. As you may know, shingles comes from the chicken pox virus and lies dormant inside your body. If you become stressed from life events, illness, or exposure to environmental toxins, shingles can surface and wreak havoc. The virus was there all the time, but only when you give it conditions in which it can thrive does it actually do so.

Legend has it that Pasteur and Bechamp argued about who was right, and that on his deathbed, Pasteur conceded. However, from my perspective, they were both correct. Yes, often there is a germ, and it needs to be treated, but if internal conditions are optimal for that pathogen, then the treatment might be ineffective. Think about a pond. If fresh water flows in and there is some drainage, the pond is a good environment for frogs, fish, and ducks to live healthy, happy lives. If the pond becomes stagnant and mosquitoes grow rampant, coming out every day with bug spray is not going to be effective. However, if you drain the pond and *then* spray for bugs, the pond will return to a healthy condition.

In my opinion, this is why we need an integrative approach whereby medical professionals treat the condition and functional medicine doctors, naturopaths, and health coaches help improve the "terrain" so that the body is not a friendly host for pathogens. So let's look at how your "pond" might need a little maintenance. To do that, we'll examine some of what I call the "good guys" and the "bad guys."

An easy way to understand this concept of good guys and bad guys is that free radicals, in excess, can be bad guys. A free radical is a molecule that has an unpaired electron. Free radicals are believed to damage cells, accelerate aging, create toxic conditions, and impair health by stealing electrons from other compounds. However, the key word is *excess,* because free radicals do play a positive role in killing pathogens; therefore, some free radicals are needed for optimal health.

As a simplified explanation, free radicals are opposed by antioxidants, the good guys that donate spare electrons and prevent free radical damage. The problem occurs when there are too many free radicals or too few antioxidants to neutralize them. If you do not have enough free radicals, you may not be able to fight off infections, but having too many free radicals can damage body tissues. This is somewhat similar to the Traditional Chinese theory of the yin and the yang: you want a healthy balance.

Although there are many free radical "bad guys," for this introduction to functional genomics, we're going to focus on just three: superoxide, peroxynitrite, and hydroxyl radicals. As for the "good guys" that neutralize the bad guys or keep them in check, we will look at superoxide dismutase (SOD), catalase, glutathione, and Nrf2, which is the transcription factor that controls the production, utilization, and recycling of these powerful antioxidants.

Before we delve into the actual genes and how they can impact excess free radicals or less-than-optimal antioxidants, let me briefly explain how genetics can make a difference between two people. In extremely simple terms, DNA is the instruction manual for how to make proteins, and particularly specific types of proteins known as *enzymes*. Enzymes take substance A, utilize cofactors (such as vitamins and minerals), and make substance B. That is how the fats, carbohydrates, and proteins we eat, the water we drink, the air we breathe, and the sunshine we absorb make every cell in our bodies and give us the energy that fuels our blood, tissues, neurotransmitters, and so on. What an incredible miracle that this is happening every second!

So DNA is our instruction manual for how to make enzymes. Sometimes we inherit instructions that cause our bodies to not make enough enzymes or to make enzymes that are not quite as efficient. When this happens, the conversion of substance A to substance B is slowed by as much as 70 percent. This can result in excess free radicals, too few antioxidants, or less-than-optimal clearing of free radicals. (Interestingly, some genetic issues speed up certain processes, but that's beyond the scope of this discussion.) We inherit this potential genetic weakness from our parents, and, depending on how the genes of our two parents combined, we can be better or worse than either parent in that one function.

The Bad Guys: Excess Free Radicals

As previously mentioned, free radicals play an important role in killing foreign invaders, but it's critical to have the proper amount. Too few or too many can be a problem. Let's examine three important biochemical pathways in which superoxide, peroxynitrite, and hydroxyl radicals have the potential to create excess free radicals that can severely hinder your body's ability to detoxify and look at how genetics and environmental factors can cause you to have too many of these free radicals. You will learn what these free radicals are, how they are made, how to reduce their production, and how to take care of them if they are present in excessive amounts. In addition to the negative impact they can have on your cells, it is harder for your body to remove toxins when it is busy fighting these free radicals.

Superoxide

Our bodies are made up of trillions of cells, and inside each cell is an "engine" called the *mitochondrion*. The "mighty mitochondria," as they are sometimes known, are powerhouses that take the fats, carbohydrates, and proteins you eat and combine them with the oxygen you breathe to make a molecule called *ATP*. In the simplest terms, ATP is the fuel that your body needs to function. If your supply of ATP were shut off, you would die immediately. Although fatigue has many causes, low ATP production as a result of mitochondrial dysfunction is often believed to be a contributing factor. Many scientific studies are looking at how mitochondrial dysfunction may be a result of exposure to environmental toxins.

When the mitochondria create ATP, they also produce a free radical called *superoxide*. Superoxide is a common reactive form of oxygen that is created when molecular oxygen gains a single electron. This free radical can attack lipids, proteins, and nucleic acids, wreaking havoc wherever it goes, and chews up valuable antioxidants and affects other mechanisms needed for detoxification.

Peroxynitrite

The 1998 Nobel Prize for Physiology and Medicine was given to three individuals for their research on nitric oxide. Although it is a simple gas consisting of just nitrogen and oxygen, nitric oxide is a powerful molecule. It has many functions, but we will limit this discussion to its impact on circulation. Quite simply, nitric oxide helps keep blood vessels dilated in order to optimize the flow of blood through those blood vessels.

You might know someone who carries nitroglycerine. If that person experiences chest pain, he or she takes a nitroglycerine pill, and it relieves the pain by boosting nitic oxide levels, which improves blood flow to the heart.

Low nitric oxide levels can cause erectile dysfunction (by decreasing blood flow), for which drugs such as Cialis and Viagra are often prescribed. To achieve an erection requires adequate blood flow. In simple terms, these drugs slow the enzymes that degrade nitric oxide so that more nitric oxide is available to support blood flow, and erectile function is improved. If a man's nitric oxide level is just a little low, these drugs can help, but if it's too low, they are not effective.

Unfortunately, there are genetic and environmental factors that can influence nitric oxide production but also make the free radical superoxide, discussed previously. A less-than-optimal nitric oxide level creates the potential for circulatory issues and high blood pressure. In addition, when superoxide is not neutralized properly by an antioxidant called *superoxide dismutase,* it combines with nitric oxide to make another damaging free radical called *peroxynitrite.* This can be a real problem.

Peroxynitrite is a contributing factor to many diseases related to inflammation, as shown in the list below. If a condition ends in "–itis," there's a good chance that peroxynitrite is part of the problem.

CONDITIONS RELATED TO PEROXYNITRITE	
Alzheimer's disease	Idiopathic pulmonary fibrosis
Arthritis	Kidney disease
Asthma	Liver disease
Atherosclerosis	Lupus erythematosus
Autism	Lyme disease
Bipolar disorder	Male infertility
Cancer	Migraines
Celiac disease	Myocardial infarction
Chronic heart failure	Neurodegenerative disorders
Chronic inflammatory disease	Osteoporosis
Circulatory shock	Parkinson's disease
COPD	Raynaud's syndrome
Depression	Skin cell proliferation and death
Diabetes	Stroke
Fibromyalgia	Ulcerative colitis

NOTE

Even if the biochem-
istry presented here
seems complicated, if
you follow along with
the descriptions, they
will help you under-
stand these important
processes and how we
can work with them to
improve health.

Nitric oxide is made when an amino acid called *L-arginine* combines with other cofactors, like BH4 and NADPH, and the nitric oxide
synthase (NOS) enzyme mixes them all together. However, many
genetic and epigenetic factors can impede this process, resulting in
what is called NOS *uncoupling*. Rather than making nitric oxide, NOS
uncoupling produces superoxide free radicals, which contribute to
the destructive path of creating peroxynitrite (see Figure 16.1).

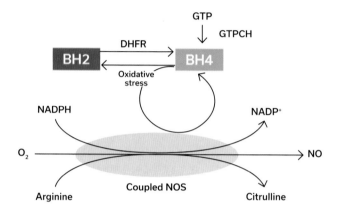

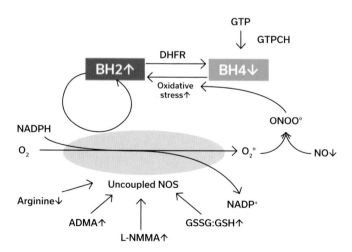

Figure 16.1. NOS uncoupling.

So what can go wrong? As previously discussed, BH4 and NADPH molecules get together to make nitric oxide. NADPH is needed for many other functions, including recycling some critical antioxidants, and BH4 helps the body make the neurotransmitters serotonin and dopamine.

Perhaps you have heard of or even taken L-arginine to support nitric oxide production. Many men have tried this to improve erectile function. If the NOS enzymes are working and there is adequate BH4 and NADPH, this treatment works beautifully. However, if genetic weakness impairs the NOS enzyme or genetic or environmental factors reduce the required cofactors, you get the superoxide free radical and possibly peroxynitrite rather than good nitric oxide. So this strategy can backfire.

Some of the genes involved in this process are MTHFR A1298 to support BH4, G6PD and ME1 to support adequate NADPH, and the SOD genes to support the production of the antioxidant superoxide dismutase, which neutralizes the superoxide free radical. If you have access to your genetic information, check to see if you have SNPs in MTHFR A1298, ME1, and SOD 1, 2, or 3, which might subject you to some difficulty with this aspect of your biochemistry.

Hydroxyl Radicals

There are many ways to detoxify, but the "master antioxidant" gluta-thione is heavily involved in supporting the detoxification process. Consequently, any free radicals that "use up" glutathione will hinder your ability to detoxify. One of the nastiest free radicals is called a *hydroxyl radical.* The bottom line is that the more hydroxyl radicals you have, the more likely you are to have difficulty detoxifying and the more prone you are to inflammation of all types.

As shown in Figure 16.2, our bodies use an intricate method to neutralize free radicals. When a free radical combines with oxygen, it creates superoxide. This on its own can be dangerous. Then we make the powerful antioxidant superoxide dismutase that turns the superoxide free radical into hydrogen peroxide (H_2O_2). Once again, the body uses H_2O_2 to kill pathogens, but in excess, H_2O_2 can cause damage. Then we need both catalase and glutathione to turn the H_2O_2 into water and oxygen (H_2O and O_2).

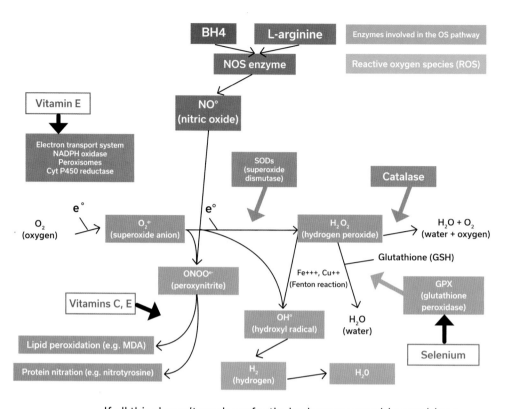

Figure 16.2. The Fenton reaction.

If all this doesn't work perfectly, hydrogen peroxide combines with iron (Fe) to make the nasty hydroxyl radicals that damage cells and DNA and make it very difficult to detoxify properly. This is called the *Fenton reaction,* a well-known chemical process discovered by a brilliant chemist, Henry Fenton, back in 1894. Fenton published a study describing how iron (II) ions, in the presence of certain oxidizing agents, yield a solution with powerful and extraordinary oxidizing capabilities. Today, Fenton's reagent is defined as a mixture of hydrogen peroxide and iron (II) ion ($H_2O_2 + Fe_2^+$).

Why might this reaction be occurring at an increased rate in some individuals? Let's see what could be going on genetically.

The problem often starts with variants in the HFE genes. Research is showing that these HFE variants are more prevalent in those of northern European descent and in Ashkenazi Jews. In simple terms, the HFE genes enable us to absorb iron from foods. Inherited genetic variants cause certain people to absorb iron at a higher rate than others. In the most severe cases, where individuals inherit a variant from both parents, there is the potential for *hemochromatosis,* a condition in which iron accumulates in the body. If not diagnosed and treated with the appropriate therapy (often in the form of bloodletting), the excess iron can cause serious damage to organs and even death.

However, even if individuals have only one variant in the HFE genes, the literature suggests that they can have slightly higher rates of iron absorption.

Interestingly, in an effort to ensure that there is enough iron in our diet, many foods are fortified with iron. So, in times of famine, the HFE variants may have been lifesaving (which possibly is why they came into existence), but when food is plentiful and enriched with iron, we may be getting too much. Excess iron can become a nasty oxidizing free radical.

Now, excess iron itself can be a problem, but it can be made worse. That's why I often refer to functional genetics as a "3D chess game played underwater." We likely won't find "the" gene that's the problem and fixing it the solution. I'm convinced that toxicity, inflammation, accelerated aging, and predisposition to disease occur in the presence of the "perfect storm": multiple genetic difficulties that contribute to detoxification difficulties and inflammation, made worse by environmental factors.

What can make excess iron worse? There are many factors involved; however, I want to emphasize the roles of ferroportin and ceruloplasmin.

The SLC40A1 gene contains instructions for making ferroportin, which plays a role in the metabolism of iron. Ferroportin is found in all cells and tissues where iron is regulated, and it is the only cellular iron exporter. Different variants of the SLC40A1 gene affect the ferroportin protein in different ways, which in turn can alter the export and metabolism of iron.

In my health coaching, I often observe unexplained and unresolved inflammation when individuals have both HFE and SLC40A1 variants. What tipped me off to the potential for this being the Fenton reaction was that many of the individuals who have this combination comment, "I've been told I'm anemic my entire life." However, when they take iron as instructed by their healthcare providers, they feel significantly worse. The reason may be that more iron is just fueling more hydroxyl radical production.

In summary, as related to excess iron, variants in the HFE and SLC40A1 genes have the potential to increase the absorption of iron and potentially set it up to be turned into hydroxyl radicals. Variants in the CP genes that make ceruloplasmin also may create conditions more favorable to iron being used to create hydroxyl radicals rather than perform its critical functions.

Having your doctor run a complete iron panel may be helpful in determining whether this is an issue for you, but keep in mind that if the Fenton reaction is turning iron into hydroxyl radicals, your iron levels may come back normal or even low.

Interestingly, there is growing interest in using hydrogen-infused water to reduce these hydroxyl radicals. It's still unclear whether a simple reaction between hydrogen and hydroxyl radicals turns the radicals into water or whether hydrogen supports the production of antioxidants, but either way, evidence suggests that hydrogen reduces the hydroxyl radicals. As a result, supplement companies are making tablets that you can drop into a bottle of water to liberate the hydrogen, and water purifiers that increase the hydrogen content in the water are coming onto the market. Although I believe that hydrogen water is as close as you can get to a fountain of youth, taking steps to slow the creation of hydroxyl radicals is just as important, if not more so. Let's explore how you can do that.

The Good Guys: The Antioxidants That Protect You and Support Detox

Superoxide, peroxynitrite, and hydroxyl radicals can damage your cells, reduce your energy levels, and make you more susceptible to illness. However, your body is capable of making the antioxidants needed to turn these damaging molecules into harmless substances that can be excreted through the urine or stool.

Superoxide Dismutase (SOD)

As mentioned previously, the body makes an antioxidant called superoxide dismutase (SOD). In simple terms, SOD takes superoxide and turns it into hydrogen peroxide (refer to Figure 16.2). Then the antioxidants catalase and glutathione and a third mechanism are needed to break the hydrogen peroxide down into harmless water and oxygen. Superoxide, if not neutralized, will create peroxynitrite, an even worse free radical.

The genes SOD1, SOD2, and SOD3 are responsible for making superoxide dismutase. They require copper, zinc, and manganese as cofactors in its production. Genetic variants, or a lack of cofactors, may result in less-than-optimal superoxide dismutase, which will cause you to age faster and be more susceptible to various health issues.

Functional genetic testing can help you determine whether you may have weakness in the SOD genes. If you do, taking steps to bolster SOD production, perhaps by supplementing with SOD, and reducing toxin exposure may be productive efforts to support the healthy function of your cells and peak energy. Some studies have shown that polyphenols from apple skins may support SOD production as well. Perhaps we now have scientific proof to validate the old saying, "An apple a day keeps the doctor away."

Catalase

Catalase is another important antioxidant made by the CAT genes. The catalase enzyme is so critical to health that it is found in nearly every living organism that is exposed to oxygen. This antioxidant can catalyze the conversion of hydrogen peroxide into water and oxygen, which alleviates the toxic effects of hydrogen peroxide.

Catalase has an extremely high rate of turnover as compared to other enzymes. One catalase enzyme can change 40 million molecules of hydrogen peroxide into water and oxygen in just one second. In fact, catalase enzymes act to protect our cells, counteracting and balancing the continual production of hydrogen peroxide.

Because of its powerful antioxidant properties, catalase is very beneficial to the organs and bodily processes. In addition to acting as a superantioxidant, catalase can use hydrogen peroxide to oxidize toxins, including methanol, ethanol, formic acid, formaldehyde, and nitrite. This type of dual activity makes it a crucial cellular enzyme.

Catalase is a key player in the body's defense against oxidative stress. Oxidative stress occurs when there is an imbalance between the production of free radicals and the body's defense against the harmful effects of those free radicals. Researchers have hypothesized that oxidative stress plays a role in the development of many chronic or late-onset conditions, such as Alzheimer's disease, asthma, diabetes, and rheumatoid arthritis.

If you have insufficient catalase due to variants in the CAT genes, your ability to clear hydrogen peroxide may be impaired. I'm sure you are beginning to realize that if you have catalase deficiencies, there is the potential for more hydrogen peroxide to collide with iron to make hydroxyl radicals, and if you have excess iron from HFE or other iron-related genetic variants or increased consumption of iron in food or via supplementation, there is the potential for the creation of hydroxyl radicals from the Fenton reaction.

Functional genetic testing can help you determine whether you have variants in your CAT genes that may impact your ability to make catalase.

Glutathione, the Master Antioxidant

Glutathione (GSH) is called the master antioxidant for good reason. Composed of three amino acids—cysteine, glycine, and glutamic acid—glutathione is found in every cell of the human body. The highest concentrations are in the skin and liver, making glutathione a critical component in the detoxification process.

Glutathione is used by the liver to clear many toxins, including formaldehyde, acetaminophen, and benzopyrene, and plays a key role in Phase I and Phase II detoxification reactions (see Chapter 15).

Glutathione is needed for the protection of proteins, which can be damaged by free radicals. It is involved in nucleic acid synthesis and plays a role in DNA repair. Glutathione maintains the cellular redox potential. Studies have shown that those people who have the highest glutathione levels live the longest. If you have insufficient glutathione, you will age prematurely and be prone to a hindered ability to detox. Individuals with low glutathione often report an inability to tolerate strong smells and are highly prone to inflammatory conditions, especially when combined with other variants that cause inflammation.

A study was done on individuals over the age of 101 who were functioning relatively well. Fascinatingly, the only commonality the researchers found was that the participants had the glutathione levels of the average fifty-year-old. Glutathione levels decrease as we age and often get very low at around age eight-five, when aging really starts to accelerate.

Making glutathione is a rather complex process because the body "assembles it" from cysteine, glycine, and glutamine. Figure 16.3 shows the enzyme responsible for each step of this process. Genetic defects, insufficient building blocks, or insufficient cofactors anywhere along this path can impede the production of glutathione. In my functional nutritional consulting, I often see multiple genetic issues in these pathways in people with serious detox issues, and medical intervention is difficult due to their poor ability to detoxify.

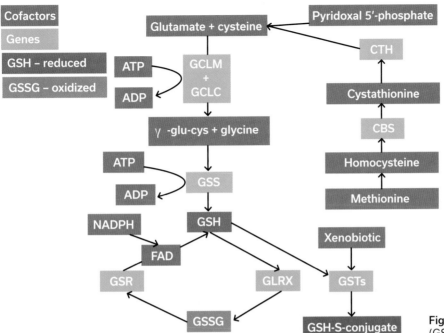

Figure 16.3. Glutathione (GSH) production.

In addition to *making* glutathione, we need to be able to recycle it after it neutralizes a free radical. When reduced glutathione neutralizes a free radical, it becomes oxidized, as shown above. The GSR gene, along with NADPH, recycles the oxidized glutathione back to its reduced form so that it can go grab another free radical.

Healthcare providers trained in functional medicine can do blood tests to measure reduced and oxidized glutathione. A decreased level of reduced glutathione, the critical measurement, could come from less-than-optimal production of glutathione or from the glutathione being used up in its efforts to rid the body of a high load of environmental toxins. Functional genetic testing can determine whether genetic variants in your enzymes needed to make glutathione may be a factor in less-than-optimal glutathione levels. This could be a key to poor detoxification.

Genetic variants in many of the glutathione genes shown above can impair glutathione production. You can compensate for this either by taking glutathione or, better yet, by supporting the production of glutathione. The most effective form of oral glutathione is liposomal glutathione, in which the glutathione is bound to a fat, allowing it to be absorbed more readily in the intestines. Perhaps even more effective is S-acetyl glutathione. However, I am of the growing opinion that taking glutathione daily, or for a long time, may eventual-ly backfire. Because glutathione is made from cysteine, glycine, and glutamine, there is the potential for these amino acids to build up in the body and have unintended consequences over time. Instead, as you will learn next, you can make efforts to support a process called Nrf2, which is how your body makes glutathione.

Nrf2, Master Control of Detox and Oxidative Stress

You have learned how excess free radicals or less-than-optimal anti-oxidants can cause you to detox poorly and age faster, which can lead to health challenges. Now we are going to move on to a process that I call the "master control" of the antioxidants, known as Nrf2 (pronounced "nerf-2").

First, an analogy: In many buildings, there is a sprinkler system. The sprinklers do nothing unless the system detects a fire. When it detects a fire, it releases water to the sprinklers to put out the flames. However, if the sprinklers are defective or there is no water to supply them, the building could burn down.

In a similar fashion, we have a "sprinkler system" in our cells known as Keap1. Through what is called *antioxidant response element (ARE)*, Nrf2 creates and releases antioxidants (the "water") to neutralize the free radicals (the "fire," so to speak). Figure 16.4 shows how Keap1 releases Nrf2 and stimulates the ARE that triggers anti-oxidants to neutralize free radicals.

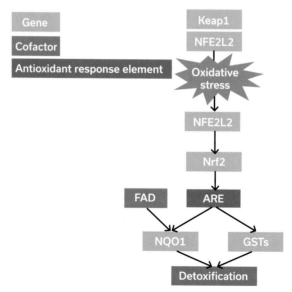

Figure 16.4. Keap1.

CHAPTER 16: How Your Genetic Makeup May Affect Your Ability to Detoxify

As you look at Figure 16.5, notice that Nrf2 controls the enzymes that make, utilize, and recycle glutathione. Of equal importance, they also are involved with the proper use of iron and the making of NADPH, which recycles glutathione (which is needed for nitric oxide production) and supports the clearing of hydrogen peroxide.

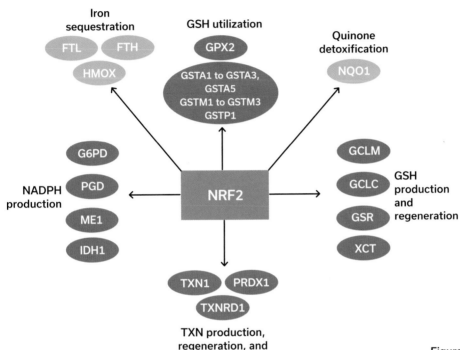

Figure 16.5. Nrf2.

Are you connecting the dots here? If Nrf2 is not working properly, your body won't handle iron properly, and it may make excess hydroxyl radicals. Without NADPH, you can't make nitric oxide effectively, and you may make more superoxide and peroxynitrite. Also, without enough NADPH, you won't be able to clear the hydrogen peroxide that combines with iron to make hydroxyl radicals. And finally, without optimal Nrf2, you won't effectively make, utilize, and recycle the master antioxidant glutathione. Now you see why optimal Nrf2 activity is so critical.

I frequently observe that those people with variants in their Nrf2 genes suffer from inflammatory conditions that no one seems to be able to resolve. Functional genetic testing can help determine whether you have issues with your Nrf2 genes.

Fortunately, there are some natural substances, like sulforaphane from broccoli, resveratrol (found in the skins of grapes and in certain berries), bacopa, milk thistle, and turmeric, that may help support healthy Nrf2 activity.

Autophagy: Another Key Component of Detox

Autophagy is another key component of detoxification. The research into autophagy earned a Nobel Prize in Medicine in 2016.

In very simple terms, autophagy is something like a janitor. It gathers up and recycles dead cells so that they can be used as fuel for new cells, and it supports the clearing of cellular debris. It accomplishes this task via an intricate process whereby the body makes autophagosomes, which envelop, clear, and recycle the old damaged cells and debris.

The ULK1, ULK2, and ATG13 genes are responsible for supporting autophagy. When individuals have variants in these genes, autophagy may be hindered.

One clue that autophagy is not working optimally is the development of what are sometimes called age spots, sun spots, or liver spots on the limbs, face, and chest. As we age, our genetic weaknesses or environmental conditions cause autophagy to slow, and these spots become more prominent.

Interestingly, there is a process called *mTOR,* which is responsible for the growth of new cells. Let's call it the "construction crew." Here's where things gets interesting: mTOR, when active, inhibits autophagy. One of the major research projects of the NutriGenetic Research Institute is to find out how environmental factors not present fifty to seventy-five years ago (like electromagnetic fields, growth hormones given to animals, pesticides, iron fortification, and MSG) are overstimulating mTOR and thus weakening autophagy. This overstimulation of mTOR and the resulting weakness of autophagy may be contributing to the rise in many serious health conditions that we are seeing today.

Functional genetic testing can help determine whether you may have some inherited weakness in autophagy.

What stimulates mTOR (the "construction crew") is "building materials," such as amino acids, iron, folate, glutamine, and carbohydrates. Consequently, during periods of caloric restriction, such as fasting or intermittent fasting (in which daily food intake is limited to a six- to eight-hour window), the construction workers take a break and tell the "janitors" (autophagy) that they can come clean up. This is likely why intermittent fasting is gaining in popularity, as we now have many new environmental factors that are overstimulating mTOR and weakening autophagy. The low-carb, low–amino acid ketogenic diet, which also slows mTOR, is gaining in popularity as well.

Finally, resveratrol and turmeric have been shown to slow mTOR, while berberine and lithium support autophagy. Taking steps to slow mTOR and boost autophagy may be a critical part of supporting healthy detoxification and supporting longevity.

So What Do You Do?

It's quite likely if you have not been exposed to functional genomics before, this material may seem overwhelming, confusing, and possibly scary. Many people are reluctant to know about their DNA, fearing that they will receive terrible news.

One of the great things about functional genomics is that, in reality, there is no bad news and nothing to be scared about. As I said at the start of this chapter, you are not defective or dirty, and you are not doomed. This is not like disease genetics, where a geneticist or medical doctor reviews genetic markers that raise the risk of some rather serious illnesses. This kind of genetic testing can show you your *potential* weak spots and give you information to make sensible lifestyle choices and options for supplementation.

The most popular genetic tests for consumers are Ancestry (www.ancestry.com) and 23andMe (www.23andme.com). Both are known for their ancestry information but also allow you to download what is called your "raw data."

There are several online services to which you can upload this raw data, and the service will tell you if you have any genetic variants. Some are color-coded, with yellow for one inherited genetic variant *(heterozygous)* and red for two inherited variants *(homozygous).* As good as these services are, many people find them quite confusing and don't know what to do with the information they receive.

Because this is an emerging field, more research needs to be done to determine whether some of these variants are clinically significant. Also, while most genetic variants are a down-regulation (a reduction of enzyme activity), other genetic variants are an up-regulation, meaning that the enzyme works faster. This is why I often refer to it as a 3D chess game played underwater! It can be rather complex.

In my opinion, we should never make decisions based solely on genetics. A commonly used phrase is "don't treat the SNPs." However, I firmly believe that SNPs can give a qualified health professional clues about where to look and which lab tests to recommend. This, combined with the typical clinical assessment, can help us learn how best to support patients' or clients' enzyme function. For example, as discussed earlier, inflammation could be from excess free radical production or from less-than-optimal antioxidant production. Knowing where the problem lies gives functional health practitioners the ability to create a personalized protocol for your unique genetic makeup. This is no longer one-size-fits-all; we are now moving toward personalized protocols to support optimal function.

To further this research, I founded the NutriGenetic Research Institute and have been passionately involved in contributing to our understanding of functional genetic variants. To date, we have performed five analyses of individuals with chronic Lyme disease and have presented the research at national and international conferences. The posters and videos can be found at www. nutrigeneticresearch.org.

Functional Supplementation

When they hear the word *supplements,* most people think of single vitamins or multivitamins. An example of when this type of supplementation may be beneficial is when the PEMT gene contains variants. These variants can create the potential for decreased levels of the B vitamin choline, which the body needs to utilize fats. Similarly, people who have variants in the PANK genes may benefit from vitamin B5 supplementation in the form of pantetheine. Finally, variants in the genes known as MTR and MTRR may indicate that supplementation with B12 could be helpful. However, as stated earlier, it's best not to "treat the SNPs," but to work with a healthcare professional to determine whether these SNPs are creating an issue for you.

Although single vitamins and multivitamins can be very valuable, they often are not enough to deal with the high load of toxic material to which we are exposed in our modern environment. Fortunately, Mother Nature has provided us with phytochemicals from the plant kingdom that have unique properties in supporting the detoxification pathways. Here are a few examples of how Mother Nature may have the potential to come to your rescue.

Perhaps your mother or grandmother encouraged you to eat your broccoli. We now have solid scientific evidence for why it is so good for us. Interestingly, broccoli contains a phytochemical called *sulforaphane* that stimulates Nrf2. As explained earlier, Nrf2 is the master control of making, utilizing, and recycling glutathione, the antioxidant needed to help remove toxins!

You likely have heard of turmeric, the spice that has been used around the world for centuries. It is gaining quite a bit of attention for its anti-inflammatory properties. Although we may never know all the ways this spice can be helpful, in addition to supporting Nrf2, it's been shown to down-regulate mTOR. I have emphasized the importance of understanding that the overstimulation of mTOR is believed to contribute to inflammation.

Another common plant that has been used for centuries is milk thistle. The herbalists of old knew that it helped the liver but had no idea why. Once again, it has been found that milk thistle supports Nrf2 and the production of glutathione, which the liver needs to perform critical detoxifying functions.

As you learned earlier, peroxynitrite is created either when the nitric oxide synthase (NOS) enzyme is not functioning optimally or when there is a lack of cofactors; as a result, this free radical is produced rather than the beneficial nitric oxide. One of the cofactors needed to make nitric oxide is BH4. Once again, Mother Nature to the rescue: royal jelly, the food fed to the queen bee, is a naturally occurring source of BH4.

If the NOS enzyme is not working as effectively as it should to make nitric oxide, nutrient-dense foods and herbs, such as green coffee beans, beet root, and hawthorn berries, have the potential to provide support. Exciting recent research has suggested that numerous traditional herbs and nutrients may support healthy NOS function.

Finally, if a "cleanup crew" is needed, Mother Nature has once again provided us with phytochemicals that appear to help "scavenge" the peroxynitrite. Fascinatingly, echinacea, garlic, rosemary, and witch hazel are showing promise that they may reduce peroxynitrite.

In summary, I believe that as chronic conditions increase as a result of environmental factors, being proactive with your health includes knowing your potential genetic weakness in the detoxification pathways, taking some commonsense approaches, and utilizing what Mother Nature has provided.

Summary

While I understand that this material is complicated, I think Bob and Matthew have done an outstanding job of simplifying these important concepts so that we can begin to integrate the information obtained from functional genomics and biochemistry into helping the sickest patients recover their health.

If you are seeking healthcare professionals who have studied this material, please visit www.yourgenomicresource.com, which also includes some excellent videos. This website can help you determine your own genomic information by providing a saliva test, which utilizes the software that Bob developed to extract the information that will be of greatest value to you. Healthcare professionals can complete an online course to enable them to work with this information on a deeper level. Additionally, the NutriGenetic Research Institute website has information and diagrams that can help you understand this material on a deeper level.

CHAPTER 17

Rebooting Multiple Chemical Sensitivities (MCS)

Fe, Fi, Fo, Fum...

Patients with multiple chemical sensitivities (MCS), also referred to as environmental illness (EI), are common in my practice. As I had the opportunity to work with a large number of patients with chemical sensitivities, I began to realize that it is a rare patient with MCS who does not also have mold toxicity. Mold toxicity appears to either cause the MCS or add significantly to the symptoms of MCS. And a large percentage of mold-toxic patients eventually develop MCS if the mold is left untreated.

The diagnosis of MCS is quite specific and readily made. If a patient reports intense symptoms *immediately* upon exposure to a scent of any kind—within minutes or even seconds—MCS is the likely culprit. Common scents that cause this reaction are cigarette and wood smoke, gasoline and other petroleum products, detergents, perfumes, and cleaning agents. You might consider MCS as being similar to mast cell activation syndrome (see Chapter 5), with the understanding that mast cell activation is triggered mostly by what is ingested (food and drink), whereas MCS is triggered mostly by inhalants.

The more common symptoms include fatigue, exhaustion, cognitive impairment, headache, neurological symptoms (such as seizures, pseudoseizures, dyskinesias, tics and spasms, syncopal episodes, and tremors), dizziness, weakness, and nausea. Once triggered, these symptoms may persist for hours or days, and they can be debilitating. As MCS progresses, it is often accompanied by anxiety, which has led many physicians to label it as yet another psychosomatic illness. It is not.

Imagine that you are walking down the street, feeling fine. You pass someone who is wearing perfume, and suddenly you feel weak, exhausted, and almost unable to continue walking. After several such attacks, you begin to worry that every passing stranger could trigger an episode that is going to set you back for days. The anxiety thus generated is understandable, and it will not go away until you have recovered completely and you can be assured that it is safe to go out in public.

It has long been argued that MCS is primarily a psychological illness, and many patients with MCS find that their symptoms are not always taken seriously by their families or their healthcare providers. Having treated several hundred sensitive patients with this illness, I am certain that it is primarily a physically triggered event. After we successfully treat the cause (most often mold toxicity or *Bartonella* infection), the MCS slowly disappears. However, the symptoms do not go away instantly; it seems to take a year or more for them to resolve completely. During that time, most patients experience gradual improvement and are delighted to discover that they can once again go to church or to the grocery store without fearing a debilitating episode.

I don't want to sound like a broken record, but again, mold toxicity and *Bartonella* top the list of conditions that set off MCS. Therefore, once a patient develops MCS, it is particularly important that he or she be evaluated for those two conditions and begin the necessary treatment. This effort helps enormously, but once MCS is triggered, other forms of treatment may be of great value in quieting down an overreactive nervous system, including many of the therapies discussed in Chapters 9 and 10 on rebooting the nervous system and the immune system.

Understanding MCS

Recent research gives us several new ways to understand MCS. We have learned that the olfactory nerves (the first cranial nerves that receive information from the sinuses and transmit that information to the olfactory lobes of the brain) can be stimulated by a wide variety of airborne particles that carry scents. This means that the nervous system can be stimulated to react within seconds of breathing in molecules that attach to nerve receptors in the sinuses and produce all the unusual phenomena seen in patients with MCS. This helps us understand the rapidity of the MCS response and see that it is not an allergic reaction, but a direct chemical reaction.

Another fascinating explanation comes from the research of Larry Klapow, PhD, who discovered a microscopic worm in the sinus and lung regions that he named *Varestrongylus klapowii (Vk)*. Research suggests that this worm may be widespread. Dr. Klapow, who suffered from ME/CFS for many years, was looking for the cause of his condition when he found this relative of the roundworm in 1991. With little funding, he has presented his findings to many medical groups, completing a study in which he demonstrated the presence of this worm in 60 percent of patients with ME/CFS and in none of the controls. I have had the privilege of working with Dr. Klapow, and we found that 95 percent of more than 150 patients with MCS were also harboring this worm in their sinuses (which we discovered by looking at material obtained from nasal washings).

Roundworms are capable of making chemicals called *acetylcholinesterase inhibitors.* Acetylcholine is a neurotransmitter whose function is to carry electrical information across a synapse (the gap between two neurons) to continue the transmission of that information to other neurons. The enzyme acetylcholinesterase helps break down that neurotransmitter after it has done its job, recycling its components so that the body can make more acetylcholine. If the function of acetylcholinesterase is interfered with and this breakdown is prevented, the transmission of nerve impulses can literally be paralyzed, which could provide an explanation for the immediate reactions to chemicals, including fatigue, cognitive impairment, and neurological disorders, that MCS patients experience.

Dr. Klapow and I were able to show in one small study that the *Vk* worm appears to make large quantities of an acetylcholinesterase inhibitor *within a minute of exposure* to a chemical known to set off an MCS reaction. We hypothesize that the chemical to which a patient is sensitive is essentially annoying to the worm that is present in the sinuses and stimulates it to make an acetylcholinesterase inhibitor that sets off these symptoms. We clearly need to do much more research in this area, but this is an intriguing explanation for the rapid onset of MCS symptoms.

> **"**
> **Having treated several hundred sensitive patients with this illness, I am certain that it is primarily a physically triggered event.**
> **"**

Rebooting MCS

Regardless of how exactly the MCS reaction is triggered, it is very real and requires treatment. MCS adds another level of reactivity to already reactive and sensitive patients. Stabilizing MCS and quieting it down is an important part of providing the proper treatment for sensitive patients. The following sections outline the rebooting methods that I have found to be the most effective.

Dynamic Neural Retraining System (DNRS)

I referred to this modality in Chapter 9 on rebooting the nervous system, but because it has been the single most effective treatment for many of my patients with MCS, I want to bring it to the forefront of this discussion as well. Developed by Annie Hopper, the DNRS program consists of visualizations and other exercises derived from recent research on neuroplasticity, which helps reboot the limbic system and quiet down these reactions.

For sensitive patients who cannot leave their homes, buying the DVDs and practicing with them is a practical way to begin. Many patients who have improved with the use of the DVDs have been able to attend seminars with Hopper that they have found to be of great value.

Low Dose Allergen Therapy (LDA)

Allergy, by definition, implies that a line of immune cells has been sensitized to react to a particular substance that binds to a receptor on the surface of the immune cell to set off a predictable reaction. MCS is less well understood, but from a practical perspective, by rebooting the T-regulatory cells using LDA, the body becomes less reactive to those substances.

While LDA has been used to treat allergies, it is less clear that MCS is an allergy; for example, allergic reactions take ten to fifteen minutes to develop, but MCS reactions can begin just seconds after exposure. When LDA was first used to treat MCS, it still was thought that MCS was some variation of an allergic reaction. It turns out that LDA is a very good treatment for MCS, but perhaps not for the reasons it was first applied.

This therapy is derived from the treatment known as *enzyme potentiated desensitization (EPD),* which was developed by Leonard McEwen, MD, in the mid-1960s. It consists of injecting tiny amounts of very diluted allergenic material just under the skin to which the immune system can respond, along with beta-glucuronidase, an enzyme that amplifies this immune response. The treatment appears to work by teaching the T-regulatory cells (which play a key role in creating the allergic response) *not to respond* to the allergen. Over time, many patients with a wide array of allergies experience a long-lasting cure.

W. A. ("Butch") Shrader, Jr., MD, modified EPD treatment into what is called low dose allergen therapy, or LDA, and has taught it to many physicians through the American Academy of Environmental Medicine. This treatment has been shown to be effective for inhalant allergies, food allergies, chemical allergies, and allergic reactions of all kinds. Some of these conditions, such as eczema and asthma, respond quickly to treatment. MCS is more complicated, and because it may not be an allergy per se, two to three years of LDA treatment with a combination of materials often is necessary to be effective on a long-term basis. I suspect that the key component here is the desensitization to specific chemicals that trigger MCS in susceptible patients. I have used LDA treatment in dozens of patients and have found it to be very helpful in reversing MCS for most of them.

A newer version of LDA is low dose immunotherapy (LDI), which is discussed in Chapter 10. I am emphasizing LDA here because of its excellent track record in treating MCS; the newer LDI has not been in use long enough for us to know whether it will be equally effective for MCS.

Nambudripad's Allergy Elimination Techniques (NAET)

NAET, also discussed in Chapter 10, was one of the first treatments available for chemically sensitive patients. Briefly, it consists of combining the principles of acupressure and kinesiology to desensitize a patient to a particular allergen. When I started seeing these patients in the early 1990s, I found NAET extremely effective in quieting down reactivity in some of them. While it is still a potentially excellent tool, I have found the DNRS program (opposite) to be faster and more broadly applicable.

Toxicant-Induced Loss of Tolerance (TILT)

TILT is a different understanding of chemical sensitivities described by Claudia Miller, MD, in 1996. She has published numerous journal articles and co-authored a book, *Chemical Exposures.* Dr. Miller proposes a three-stage process in the development of chemical sensitivities, starting with an initial chemical exposure that causes a breakdown in tolerance. As toxins accumulate, the body reaches a "tipping point" that is followed by the onset of a wide variety of symptoms caused by low levels of daily exposure to chemicals, foods, and drugs. Following this, a "masking" stage occurs that obscures the relationship of the triggers to the symptoms. Dr. Miller sees a great need for treatment centers that can separate patients from their usual environmental triggers and allow physicians to identify with precision the materials that are creating the persistent symptoms.

To help sort this out, Dr. Miller utilizes a questionnaire, the QEESI (Quick Environmental Exposure and Sensitivity Inventory), which can be readily accessed on her website: drclaudiamiller.com. The essence of treatment is to identify the toxins that are affecting the body and remove them. Mold toxicity is included in the list of major exposures. The TILT concept is quite compatible with our discussion of how to treat chemically sensitive patients.

There are other techniques available for treating MCS, but the ones discussed here have worked best for my patients. The take-home message of this chapter is that MCS commonly accompanies the illnesses that appear to create extreme sensitivity (including mold toxicity and Lyme disease and its co-infections), so diagnosing and treating those illnesses is paramount. Treating MCS concurrently with those illnesses is very helpful in settling down a highly reactive system so that patients can embrace the full treatment programs being provided.

CHAPTER 18

Rebooting How We Handle Stress

"I'm Freaking Out": Agitation, Anxiety, Depression, and Despair

To suggest that stress is a major problem for anyone with a chronic illness is to belabor the obvious. Of course it is! Merely having an illness brings with it profound effects not only on you, but on everyone you relate to. Not being able to count on functioning normally creates all kinds of disturbances. You cannot make plans, or, if you do, you cannot always carry them out. If you begin an activity, you may have to curtail that activity prematurely. This leads to disappointment for you and for your family and friends. You never know if you will have the energy or mental focus needed to begin or complete a designated task. You never know if your pain will be a limiting factor or if your sensitivities to light, sound, foods, or chemicals will stop you in your tracks. As your illness persists, you begin to wonder, *Will I ever get better? Can anyone help me?* At times, you become so anxious that you want to crawl out of your skin. And at times, hopelessness and despair take over.

All too often, your family and friends do not understand what is happening to you. You look fine on the outside but feel terrible on the inside. You are in a double-bind as to how to handle the situation. If you keep describing to others how awful you feel, you run the risk of alienating everyone you talk to. On the other hand, if you are stoic and don't talk about what you are experiencing, no one will have the slightest idea about the severity of your condition.

There is very little middle ground here. If your doctors keep telling you that your illness is "not real" and give you no clear diagnosis to work with, then your family and friends will have little recourse but to believe those medical professionals. Then you are truly alone and suffering, with no one believing that you are really sick. This dynamic adds exponentially to the difficulties you are wrestling with, which are hard enough to cope with even if you have adequate support.

Throughout the course of illness, many of my patients are challenged by intense, at times overwhelming emotions. As upsetting as these experiences are in and of themselves, *the way they are handled* by medical professionals, family members, and friends often exacerbates these experiences to the point that it adds a PTSD (posttraumatic stress disorder) component to an already difficult situation.

Far too often, the conventional medical approach has been to quickly label the patient with (take your pick) anxiety, depression, OCD, "the spectrum," bipolar disorder, or even schizophrenia or being flat-out "crazy." The next logical step is to refer the patient to a psychiatrist for immediate treatment with medication. Once this step is taken, the search for the cause of these emotional symptoms is abandoned, and the unfortunate patient is left up the metaphorical stream without a paddle. As I hope you realize in reading this book, looking at these experiences in this way and treating them purely symptomatically without seeking the cause is not helpful in either the short term or the long term.

While there are many books devoted to coping with stress, I would like to outline two major strategies for working with symptoms such as anxiety, depression, and mood swings using our rebooting model.

Understanding the Cause(s) Makes Symptoms Far Less Frightening

Helping patients get to the root cause of their symptoms goes a long way toward helping them cope. There are, invariably, physical components to these emotional symptoms.

As emotions burst forth, the fear is, *"I am going crazy—I'm having a mental breakdown—I'm freaking out"* and, ultimately, *"I am losing my mind."* Could anything be more terrifying? I find that it helps to explain to patients that they are not going crazy from some deep-seated mental disorder over which they have no control, but rather that their bodies are wrestling with a surge of toxins and/or inflammation. The toxins are usually from a microbial source, typically mold, the bacteria *Bartonella*, or the parasite *Babesia*. Knowing that the issue is *physical, not mental, in origin* makes a huge difference for patients psychologically. In fact, I have taken to describing these symptoms as "physical anxiety," and this explanation immediately resonates with most of my patients as a clearer and less frightening way to understand what is happening to them.

As discussed in Chapter 3, Dr. Ritchie Shoemaker published a helpful paper clearly showing that mold toxins (and Lyme toxins, in a different way) specifically affect the area of the brain in which emotions are processed, called the amygdala. Changes in the structure of the amygdala, measured on a special MRI called a NeuroQuant, can be reversed with the proper treatment of mold toxicity, with a concomitant improvement in all the symptoms of mold toxicity. *What this means, in simple English, is that the symptoms of mold toxicity, even those that appear to be psychological in nature, are physical and treatable.*

When they understand this physical component, my patients can remind themselves that their reactions are, indeed, real and not "all in their heads." They can remind themselves that the waves of emotions they are experiencing are caused by fluctuations in the levels of toxin in their bodies and that these waves will pass. Having this information in mind is not only comforting but also gives patients a tool to help them cope with symptoms as they arise. When the waves pass, the patients receive additional confirmation that these exacerbations are temporary.

If you don't mind a small digression here, I would like to provide yet another explanation for these kinds of psychological episodes. This point is to help you understand that, to a certain extent, these phenomena are created by the consciousness of the provoking microbes.

I know it strains the limits of credulity to bring up the consciousness of a single-celled organism. There is, however, a growing body of scientific research demonstrating that some of the microbes living inside our bodies do have, for want of a better word, "consciousness." Perhaps "hive consciousness" is a better way to put it.

The most obvious example is how the yeast *Candida* creates cravings for sweets or carbohydrates in those who are harboring it, especially in the gut. The hive mentality of *Candida* might be thought of as being similar to the Venus flytrap plant in the movie *Little Shop of Horrors* that demands, "Feed me, Seymour!" This collection of yeast cells makes known to you its intense desire for sugar, and with this imperative coming from within your own body, you assume that you (not the yeast) are the one having this craving. Now, I don't know about you, but I think of myself as just "Neil," not as Neil and his 70 trillion bacterial friends that live in his GI tract. Therefore, it does not often occur to me that those microbes might have needs of their own and might actually express those needs. Whenever "my" body (is it my body or theirs?) tells me I need something, I assume I am listening to my own personal needs. You can see how the message can be easily misinterpreted. Honoring this craving for sweets might make the *Candida* happy, but I'm not sure it would do much for me. In fact, I would essentially be feeding the *Candida,* which obviously would *not* be good for me in the long run. A high-protein, low-carb diet that omits sweets is a major component of treatment for exactly this reason. (See Chapter 12 for more on the nutritional component of treatment.)

In a similar way, the "hive consciousness" of *Bartonella* communicates how desperate and hopeless it feels as it is successfully treated (and is dying), and patients mistakenly believe that these emotions are their own, and not those of the microbes residing inside their bodies. As farfetched as this theory might sound, every patient with whom I have discussed it has responded with something like, "That makes sense; that's how it feels to me." Their experience (which is correct) is that these intense emotions are *not theirs*. Rather, the source is external (technically, internal) in the sense that these symptoms are coming from the dying bacteria and not from their own minds. This explanation helps patients process these feelings differently and cope with emotional swings much better.

In this context, the use of medications and supplements for the treatment of emotional symptoms can be helpful, but only as an adjunct to more comprehensive treatment of the underlying cause. They will not work as stand-alone therapies. *I cannot emphasize enough that treating the cause of these symptoms has to be the primary goal.*

Having said that, there is a wide variety of modalities that can be helpful here. Being a purist about getting to the cause is great, but my patients need help dealing with the vicissitudes of their emotional swings so that they can function at the highest possible level while undergoing treatment.

- Many of my patients find that tiny doses of benzodiazepines (for example, Xanax, Valium, Ativan, or Klonopin) or SSRIs (selective serotonin reuptake inhibitors, such as Celexa, Lexapro, Prozac, or Zoloft) help them through these emotional roller-coaster rides. As little as one-quarter of a 0.25-milligram Xanax tablet might be sufficient.

- Stephen Buhner has written extensively about herbal approaches (see his books on *Bartonella* and *Babesia*), especially noting the benefits of Chinese skullcap root, *Pulsatilla,* greater celandine, and motherwort. Valerian, kava kava, and passionflower have a long history of quieting the nervous system as well as enhancing sleep.

- The homeopathic remedies PSY-stabil, Viscum, and aconite have been particularly helpful for many of my patients as well. For sleep, the homeopathic Somcupin has been beneficial, as have 5-HTP (5-hydroxy tryptophan), L-tyrosine, Fibroboost, and GABA.

- The brain has cannabinoid receptors, meaning that it has evolved a relationship with cannabinoids over the centuries and responds to them with great specificity. When patients are experiencing debilitating anxiety, pain, nausea, and/or insomnia, I find the use of cannabinoids to be of great value. While each state has its own laws governing cannabinoids, the use of the cannabidiol (CBD) strain allows patients to obtain the medical benefits without concerns about getting high. CBD oil is now available to almost everyone for medicinal purposes and can be used topically, orally, or by inhalation.

Working with Thoughts and Feelings Directly

The second major helpful approach to rebooting these emotional upheavals is to work with thoughts and feelings directly. It is not commonly appreciated that these intense emotions and thoughts (*"I'm freaking out"*) can be worked with and treated by using a wide variety of meditative practices.

Thoughts are, intrinsically, fleeting experiences. The practice of meditation (prayer, Vipassana, Zen, mindfulness, yoga, and so on) teaches that if you quiet your mind, you can learn to observe that thoughts occur almost as bubbles floating up and then dispersing. Simply observing this process allows us to perceive a never-ending sequence of "thought bubbles." Some people think (erroneously) that the purpose of meditation is to learn how to control these thoughts and shut them down. That is not possible; thoughts will arise and bubble through, and then new ones will take their place. That's how the mind works. We can't shut down this process (and it probably isn't healthy to try), but we can learn to step back from those thoughts and just observe them going by.

The problem for most of us is that we get hooked on or attached to particular thoughts and can't let them go. Then they control us, and we feel helpless to do anything about it. Rather than let a thought go, we cling to it and then are held within its grasp. It is the clinging that is the problem. For example, let's take the thought *"I'm freaking out."* It is just a thought. But if you are frightened by that thought and grab on to it, it can linger for a long time. It may gain power with each passing moment until it literally takes over.

The good news is that there are several ways to reverse this process. One is to focus on your breathing to distract you from the hold that this thought has achieved over your consciousness. As you get distracted, the thought loses its power and finally fades away, which is its natural tendency. Virtually all schools of meditation use breathing as a tool to accomplish this distraction process. As you continue to study, it changes from a distraction process to a focusing process that enables you to gain more perspective and get less and less attached to thoughts as they float past.

> **"**
> **The problem for most of us is that we get hooked on or attached to particular thoughts and can't let them go.**
> **"**

Though the concept is simple, getting good at it does take a bit of time. And simple it is. You just sit quietly and comfortably and intently observe your own breathing. You notice how the air feels as it moves through your nostrils, down your windpipe, and into your lungs; then it pauses; and then you notice how it moves in the reverse direction, pauses again, and repeats. By taking a few minutes to perform this exercise, you will quickly realize that these intense thoughts that have taken over your consciousness have begun to dissipate. You regain control of your own mind. No longer does your mind run you.

Once you have gotten the hang of observing your thoughts and letting them bubble away, you can go a little deeper to ask, "Who or what is thinking these thoughts?" This process is described very nicely in the book *The Natural Bliss of Being* by Jackson Peterson. He gives the example of taking a moment to picture a dog in your mind. Then you step back and notice what part of you is doing the picturing. As you do so, you realize that the picture of the dog fades quickly as you get in touch with this part of you referred to as the "observer."

Learning how to step back from what you are imagining or thinking by using meditative techniques is a powerful way to pull yourself away from a particularly difficult emotion or thought and regain equilibrium. Some of the better resources for getting started include the following:

- Jack Kornfield's *The Inner Art of Meditation* CD, which takes listeners through the process step by step (and Jack is a great storyteller, so his material is easy to listen to and absorb).

- The website Headspace (www.headspace.com), which provides information about meditation in easy-to-digest increments.

- I also like the mindful approach to yoga provided by Yoga with Adriene (www.yogawithadriene.com), which can be gentle and done in brief increments. This approach enables you to develop an awareness of exactly where you are holding tension in your body and use specific stretching exercises to release that tension.

- Another method that has helped many of my patients is Heart-Math, which is essentially a biofeedback system that displays your heart rate on a device (such as your phone) and enables you to shift your heart rate variability (HRV) to a more synchronous pattern. Demonstrating a more coherent physiological state not only can decrease stress but can create a more balanced emotional and cognitive state as well.

My message here is simple: Please take the time to learn and practice a method that resonates with you to quiet the emotional turmoil inside so that you can cope better with everything on your plate and move toward healing. If you are dealing with family and friends who cannot comprehend your illness and are not supportive of you, the best solution is to use the information and treatments outlined in this book to get well. They may never fully understand, but if you get well again, it won't matter.

CHAPTER 19

Rebooting the Psyche: Emotional Awakening

Because the body, mind, and spirit are intricately interwoven, it is impossible for anyone who is ill not to suffer emotionally as well as physically. I cannot think of any physical illness that does not create worry or fear, which only adds to the burden of suffering. In patients who are unusually sensitive or toxic, this suffering is magnified many times by the direct effect of those toxins and/ or infections on the nervous system and brain. Mold toxins cause inflammation in the areas of the brain that process anxiety, depression, and mood swings. *Bartonella* and other infections can do the same. So can mast cell activation, porphyria, and multiple chemical sensitivities (MCS). It can be very difficult to tease apart how much of a person's emotional upheaval is due to the underlying illness and how much is due to his or her reaction to, or ability to cope with, that illness.

The only way I know to deal with this is to treat all the known causes of a patient's symptoms and, like peeling an onion, remove the physical components layer by layer and see what lies underneath each one. For some patients, simply removing the physical trigger—the mold toxin or infection—is sufficient for them to recover completely, and their anxiety and/or depression appear to vanish. For others, even when the physical triggers are seemingly gone, a large emotional overlay remains. This latter group of patients may require a rebooting of their emotional equilibrium and may benefit from the time-honored treatments of psychotherapy, hypnosis, meditation, and a wide variety of emotional release therapies.

Dysfunctional Families and Chronic Illness

One of the most difficult types of patients to treat is a patient with underlying family dysfunction that interferes with attempts at healing. What stands out for me as one of the hardest systems to address is a particular dynamic common between young women and their mothers (although we do see it in men to a lesser extent), in which Mom takes her suffering daughter under her wing and insists on directing the healing process. To a certain extent, this approach is healthy; of course a mother should do everything possible for her ailing daughter. However, in many cases we find that the mother has a deep-seated need to be needed. This need can be so intense that the family will accede to this powerful emotional demand without questioning it or even realizing that it is present. The results can range from difficult to catastrophic.

This dynamic usually polarizes the family: the mother perceives herself to be the daughter's savior, while the father is perceived as distant and unable to contribute to the girl's care. And, to a certain extent, this is true: how could a father, often trying hard to make enough money to support the family and cover the added expenses of having an ill child, begin to compete with the time and effort expended by the mother? Caring becomes the mother's crusade, and she alone carries the torch. This tends to create an intense bond between mother and daughter, leaving the father separated and confused and, as the situation evolves, angry and isolated.

Now we have a potentially unhealthy bond between mother and daughter in which the daughter "must" remain ill so that her mother can feel needed. If the daughter does improve with treatment, at a certain point she may perceive that this intense relationship is being threatened. Something will happen to sabotage the therapeutic improvement, and things will begin to backslide.

I found a clear description of this dynamic in, of all places, *The Cruelest Month,* a novel by Louise Penny, in which a psychologist describes the difference between love and attachment. The psychologist explains that love is found when the person who is loved is allowed to individuate and grow and optimize his or her gifts and talents. Attachment is seen as smothering to the point that a hostage-type situation is occurring. Similar to the well-described Stockholm syndrome, the one who is smothered, or taken hostage, develops a love-hate relationship with the captor and becomes simultaneously resentful and defensive of that relationship.

The effects created by this dynamic are clearly problematic, but because of the intensity of the emotional attachments manifest in this situation, this subject is difficult for the treating physician to address. The mother responds to any attempt to clarify this dynamic with this type of response: "Are you questioning my devotion to my daughter? Am I not doing everything humanly possible to help her? I am devoting my entire life to this effort; are you telling me that a mother should not do everything in her power to help her daughter who is suffering mightily? Shame on you!" The daughter's response is: "How dare you criticize this saint who is my mother! Don't you realize what she has sacrificed for me? Shame on you!"

Now we are at a real impasse. In her legitimate, well-intentioned efforts to help, Mom has painted herself and her daughter into a corner of their own making, and any effort to challenge that scenario is viewed as heresy.

A surprising number of the sickest patients I have treated could be described by this relationship—unfortunately, it is not rare. Unless the daughter can be removed from this family dynamic, no progress can be made. However, when the daughter can be separated from her family so that she can begin to see the truth of the situation, I have witnessed amazingly rapid progress. I wish I could find a way to address this issue earlier in treatment, since otherwise it tends to become a downward spiral.

What I have just described is an extreme example of how emotional issues can derail the healing process. The openness of each patient to look at this area is a good barometer for his or her likely success at healing. Those who understand their psychological issues are often the ones whose healing is the most complete. All I can ask of each patient is to be willing to look inside (a terrifying journey for many of us) to find whatever impediments to healing they can.

It is a rare human being who escapes childhood with all needs perfectly met. The vast majority of us have all sorts of residual emotional pain. I do not believe that we need to spend adulthood wallowing in this material, but I do believe that if certain core beliefs are brought to the fore, we are capable of moving past them and making our lives far more comfortable, and we will be far less restricted by old experiences and habits. Many of us learn that we have a choice—a potentially healthier choice. Though admittedly difficult, recognizing this is an integral part of the healing process.

Emotional Release Therapies

There are many, many approaches available in this realm; I will discuss the ones with which I have had the most experience and which I have seen create profound shifts in patients' emotional and physical well-being.

Reichian Therapy

In his classic book *Character Analysis,* published in 1933, Wilhelm Reich outlines a technique of evaluating breathing and muscle tightness that helps us understand how patients hold in emotions. He recognized that patients tighten up certain areas of their bodies to prevent the release of emotions when they do not feel safe, or ready, to do so. He also recognized that emotions are held in the myofascial tissues (not just in the mind), and that by working with those tissues and using breathing techniques at the same time, he could help patients access and release those emotions.

I was fortunate enough to work with Philip Curcuruto, DC, who studied directly with Reich, to learn how to implement this information. In the early years of my medical practice, I was able to help a number of patients with autoimmune illnesses and chronic pain syndromes get in touch with repressed emotions, release those emotions, and experience a complete remission of illness. (I still utilize these principles daily, but the precise technique and discipline is too time-consuming to offer now.)

Hypnosis

I took my first elective course in hypnosis while still in my first year of medical school. This was my entry into understanding the importance of precise language to instill in patients hope for the possibility of change. After hearing calming phrases and using breathing techniques to create a relaxed state, people are far more open to suggestions about how change can be facilitated. By using hypnosis techniques that they were readily able to master, I was able to assist patients with severe asthma attacks normalize their breathing in just minutes and help seizure patients prevent and control their seizures. When patients are aware that there is something "buried" in their subconscious that they are unable to access, hypnosis is a wonderful tool that can ease them into awareness so that they can move forward.

Ketamine

From my friend and colleague Joel Friedman, MD, comes this fascinating description of another method for rebooting in the psychological arena: the use of ketamine.

• • • •

We are experiencing a resurgence of interest in the use of sub-anesthetic doses of ketamine, which until recently had been used primarily for surgical anesthesia. We have Yale Medical School largely to thank for this, as they began looking into it as a treatment for suicidal ideation, and now it's used as a treatment for PTSD and depression. Most of us in the medical world know that we have virtually no effective treatment for suicidal patients, yet Yale was finding an immediate effect. This caught my attention, and not long afterward (actually, the very next day after I read the article), a young man presented to my office with suicidal thoughts as his main complaint. I asked him to rate his level of suicidal ideation on a scale of one to ten, ten being "pass me the gun." He responded, "Nine." He was not expecting a smile from his doctor when he described his distress and asked why I was grinning. I replied, "Funny that you should come today. Yesterday, I would have had nothing to offer you. Today, I have a very interesting and hopeful treatment." I went on to explain what few details I had at that time about ketamine and asked if he was interested. "Of course I am," he replied.

He lay down on my examining table, which is larger than the typical examining table and quite comfortable. I gave him headphones to block out ambient noise and an eye mask to block out light. I then gave him 35 milligrams of ketamine intravenously as a rapid infusion. I sat and watched, not knowing what to expect. He was completely still, and his breathing became slow and full. Outwardly, not much happened. He slowly came around after twenty minutes and took off the headphones and eye mask. The first word out of his mouth was "Wow," which, as it turns out, is pretty much everyone's first reaction. We spoke a bit afterward, but finding words was not easy for him. It became clear that this experience defies description, and this effect seems fairly universal; I will get to it in a bit.

I asked the patient to rate his level of suicidal ideation after the ketamine, hoping it was miraculously reduced to a six or so. He looked puzzled and responded, "Zero." I asked again, thinking he did not understand my question. He repeated, "Zero." Now it was my turn to be speechless. Going from a nine to a zero in twenty minutes, in terms of wanting to kill oneself, is, simply put, a miracle.

I had never seen or heard of this before, nor had any of my colleagues. I knew we were on to something very special, and I have been doing ketamine sessions ever since for conditions including depression, PTSD, bipolar illness, and anxiety. The only mental conditions for which I will not use ketamine are schizophrenia and borderline personality disorder; this is not a treatment for anyone who does not have a decent grip on reality. I screen every patient carefully and will not use ketamine if I feel uncomfortable about proceeding. The results have been close to universally beneficial. The level of benefit ranges from mild to life-changing. Sometimes, just one session is needed, and for other patients, it is a weekly treatment.

The mechanics of a session are straightforward. I administer ketamine in three ways: quick intravenous (IV) push, slow IV drip, and intramuscular injection. For first-time patients, I use a 35-milligram IV push. The slow IV drip is administered over thirty minutes, and the dose varies from 50 to 80 milligrams. The intramuscular dose of 50 to 60 milligrams is given in the deltoid muscle. These have proven to be very safe dosages requiring no monitoring devices, and no patient has ever experienced difficulties in terms of dosage.

It is important to discuss the ketamine experience in terms patients can understand. First I explain that the experience is very strong and unique and will be unlike any they have ever had; going into it with a degree of comfort is essential. When approaching a new experience, it is natural to be a bit anxious. This is an experience that cannot be controlled, and patients with control issues have a hard time letting go to the extent required. If a patient conveys an inability or unwillingness to surrender control for the duration of the session, then I back away from using ketamine until they feel they can let go and let the experience unfold.

It became clear to me that I needed to experience a ketamine session myself so I knew firsthand what the experience is like. The particulars of its effects on neurotransmitters and more generally on the brain are of little interest to me, but I am happy that others are looking into this aspect. I have always been interested in its practical applications, as this is the essence and focus of my clinical practice.

Within a minute of being injected with ketamine, a very odd sensation occurs. Many people describe it as an out-of-body experience (OBE), and it is. It happens with such abruptness that the patient needs to be prepared for it. It is important to recognize what is happening and to let go as much and as deeply as possible. It can feel like dying, and prior to taking ketamine, the patient must understand that he or she is not dying, even though it may feel that way. All bodily functions continue to operate normally; conscious awareness is not needed for the body to function during a session.

Others experience their OBE as a great liberation, as if they have come home. These are the ones who don't want to come back. After the abrupt OBE, what follows is an experience that varies from person to person and for one individual over multiple sessions. No two sessions are alike. Some are dark, and others are very light. Some sessions start out dark and then become very light. (I have never seen it the other way around.) Sometimes, there is a story; other times, a sequence of scenes or geometric shapes or bright colors. Sometimes the feeling is blissful and expansive; other times it can seem as if nothing happened. On questioning the latter type of non-experience, there will always be some nugget to be extracted with a little digging and reflection. Dark experiences are often followed by blissful ones, and the point must be made that there is as much to gain from one as the other.

After a session, it becomes clear that this experience is beyond the realm of words. It is unnecessary to talk afterward, but talking sometimes serves to bring a patient out of the "post-ketamine" state of mind. It takes a good week, sometimes more, to integrate the experience. I always allow the patient to decide whether more sessions are needed. Everyone seems to know if and when repeat sessions are needed and when his or her time with ketamine is done.

I feel that careful attention to the environment and setting is critical to a successful outcome. A thorough explanation of what to expect must be accompanied by a safe, quiet, and serene setting for treatment. I prefer a more natural setting over a clinical setting for a ketamine session. I find all the monitoring devices and high-tech approaches used by some practitioners to be unnecessary, and they only add to the expense of treatment.

What exactly happens during a ketamine session? Two main events take place. From a brain perspective, one experiences a "reset." This word comes from my patients. This reset can allow a shift in the basic mood or affect, the default setting in the brain. If the patient is chronically depressed, he or she can see a change to less time being depressed, or to no depression at all. If the patient has PTSD, he or she can see that trauma vanish, as if an eraser had done its job on the part of the brain where that particular traumatic memory was stored. This can happen in a single session, but some patients choose to do multiple sessions, each time getting the reset they so fervently desire and need. Responses vary among individuals in regard to the duration of benefit. One gets an antidepressant effect and a reset with each session, and the effect seems to be cumulative. I know of no other modality that even approaches the profundity of ketamine's capability to reset the brain.

While the brain reset is so beneficial and therapeutic, there is a parallel phenomenon that is more intriguing and profound: the OBE that one experiences. Contextualizing this experience is important. One feels a complete loss of sense of self. That is, there is a temporary (fifteen- to thirty-minute) loss of the mind/body, which we call the ego. A permanent loss of ego we know as death, but under the influence of ketamine, we get a sneak preview of what it will be like when we die. Most patients leave with a much more relaxed and less fearful attitude toward death, perceiving that there is a realm to which we return that seems to be eternal. This alone is worth the price of admission.

After receiving this treatment, most of my patients leave the office with a profound sense of peace and calm. In this anxiety-ridden age, what could be better or more profound?

• • • •

Testing for Nutritional and Biochemical Imbalances

Some common nutritional and biochemical imbalances can contribute greatly to anxiety, depression, OCD, ADD, ADHD, and mood disturbances. It is rare for a psychiatrist to be aware of this information, as the field of psychiatry has evolved into one that focuses on the use of pharmaceuticals in treatment. As a central thesis of this book, treating symptoms without looking for the possible causes is not a logical or effective way to practice medicine. (Thank you, Mr. Spock.) So let's look at some of these imbalances.

- **Pyroluria:** Porphyrins (discussed in Chapter 6) are the break-down products of heme that build up when hemoglobin is not recycled properly. You can appreciate the magnitude of this recycling by realizing that every second, our bodies have to deal with the anticipated natural death of more than 2.5 million red blood cells. That's a lot of heme to process. Quite a few medical observers have noted that some chronic illnesses, especially Lyme disease and mold toxicity, interfere with heme metabolism and predispose patients to this buildup of toxic pyrroles. A simple twenty-four-hour urine test can measure for the presence of excess kryptopyrroles (KPU) and hydroxyhemopyrrolin-2-one (HPL); if they are elevated, they may be a contributing factor to the emotional symptoms noted above. The lab that I rely on most for this testing is Health Diagnostics and Research Institute. The good news is that pyroluria can be treated with appropriate dosages of zinc and vitamin B6, which can lead to improvements in emotional symptoms while the causes of these imbalances are being addressed. By adjusting the dosages of zinc and vitamin B6 (and its metabolic cousin P5P, or pyridoxal-5-phosphate) and tracking urine levels of KPU and HPL, physicians can monitor and treat this condition with some degree of precision, and by doing so can substantially decrease a patient's anxiety and depression.

- **Zinc and copper balance:** William Walsh, PhD, discusses KPU as well as zinc/copper and methylation chemistry in detail in his excellent book, *Nutrient Power: Heal Your Biochemistry and Heal Your Brain.* He recommends measuring plasma zinc and serum copper levels and looking at the ratio between them. In my patients, it is common to see low zinc and relatively high copper, which can readily be reversed by supplementing with additional zinc. While some providers focus on excess copper, it is rare in my patients.

- **Methylation:** The earlier chapter on methylation focused on the difficulties that unusually sensitive patients have with taking the supplements normally recommended for treatment, which include a low dose (200 micrograms) of methyltetrahydrofolate, 500 micrograms to 2 milligrams of hydroxocobalamin (hydroxo-B12) taken daily as a sublingual lozenge, and 500 milligrams of phosphatidyl serine. The vast majority of my patients are unable to methylate properly. This is sometimes termed *undermethylation,* which was the focus of Chapter 14. Among patients with primarily psychological diagnoses, whom Dr. Walsh describes in his book, some *overmethylate,* which means that the usual treatments may backfire. Dr. Walsh defines overmethylation as a lowered red blood cell (RBC) histamine level (as measured by a test available from LabCorp). Methylation is required to remove excess histamine from the body, so a low RBC histamine level implies that this metabolic process has become overactive. In the majority of my patients with chronic illness, the opposite is the case. High RBC histamine levels suggest that methylation is compromised.

Helpful Supplements for Sensitive Patients

In addition to the nutrients discussed above, there are a few materials and treatments that I have found especially helpful for quieting down a reactive nervous system (which is directly connected to emotional distress by inflammation of the limbic system). The following are my favorites; other practitioners will have their own.

■ Chinese skullcap root *(Scutellaria baicalensis),* taken in tincture form, has been quite helpful in removing inflammation from the brain and nerve tissues. Sensitive patients should start with one drop once a day and slowly increase to one drop twice a day, then to two drops twice a day, working up to five to ten drops two or three times a day as tolerated.

■ Viscum, a homeopathic made by Pekana, can provide similar benefits to skullcap. The dosages for sensitive patients are the same.

■ PSY-stabil, another homeopathic made by Pekana, is especially good for relieving anxiety. Start with similar dosages as above and slowly work up to five to ten drops two to three times a day.

■ Aconite 30c, one pellet taken sublingually two or three times a day, is a homeopathic that can be helpful for anxiety.

■ *Pulsatilla* (pasqueflower), five to ten drops hourly, can be helpful for relieving a panic attack. Up to six doses can be taken.

■ Greater celandine *(Chelidonium),* five to ten drops two to three times a day.

■ CBD oil, if available (see page 275).

Summary

It is easy for people to get caught up in dysfunctional patterns of behavior even when they are relatively well. Illness magnifies these difficulties, and then they become an integral part of the problem. It is much harder to change established behaviors than to create new ones. As we age, change becomes increasingly difficult for most of us to accept. This chapter is intended to bring this information into consciousness so that these dysfunctional patterns can be dealt with. While I have mentioned only a few of the options available to patients, the key point is that a lot of help is out there, but patients must be willing to embark on the difficult journey to embrace that help.

CHAPTER 20

Rebooting the Spirit: The Possibility of Spiritual Awakening

So, after years of intense suffering, is there a point to it all? Is there meaning? Or are you to shake your fist at God for putting you through all this misery? That's your choice. There is an old joke in which the sufferer rails to the heavens, "Why me, God?" Surprisingly, God responds, "Why *not* you?"

After working with thousands of chronically ill patients for more than forty years, I have noticed a fascinating pattern. As patients begin to recover, some discover that they cannot imagine a life without illness. Like an old, worn-out shirt, it is so comfortable that they just can't seem to throw it away, regardless of how shabby it has become. Despite how long I've worked with this population, it always comes as a surprise when a patient who is almost well suddenly realizes that being well would require too great a change, then finds a way to sabotage healing and backslides.

That sabotage can take any number of forms, including abruptly stopping treatments that were helping, subconsciously overdoing it to the point of exhaustion, and taking on stresses that are guaranteed to be overly taxing. After all that hard work, what he or she sees as the responsibilities of being well are simply overwhelming. For a younger person, this might mean completing their degrees and launching into the workforce. For an older person, it might mean rejoining the workforce and resuming the stressful life that pushed him or her over the edge.

As disappointing as this reaction is for me and for the patients' loved ones, I have come to understand it as a spiritual crossroads, or what St. John of the Cross taught as the "dark night of the soul."

> **We must never lose sight of the goal: becoming completely well in body, mind, *and* spirit.**

I have learned not to underestimate this moment, and I try to prepare my patients for it long before it arrives. As patients improve, there comes a time when I must ask, "What will you do when you are well?" This question cannot be asked casually; it must be asked with great attention to how open the patients are to receiving it and musing upon it *without demanding an immediate response.* It is crucial not to engage any negativity with this question, but merely to broach the subject and leave patients open to pondering it in their own way, in their own time. When they appear open to further discussion, I can ask the question again and provide encouragement for the ideas they are considering. This is always phrased with a positive visualization, such as, "Wouldn't it be great to be able to plan interesting outings and know that you will be able to follow through and enjoy them?" The point is to prepare them for a return to health and vitality—and this is no simple matter. Often it involves rearranging their entire lives.

So, throughout this entire complicated diagnostic and treatment program, we must never lose sight of the goal: becoming completely well in body, mind, *and* spirit.

Embracing a New Perspective

I have consistently observed that the patients who eventually can embrace what they perceive as the spiritual messages from their illness not only recover completely but go on to be happier and more productive than they had ever been previously.

The cancer patient who is finally declared in remission now lives with the knowledge that every moment, every relationship, every smile, every lovely meal, every flowering bush, and every sunset is precious. For those who can sustain this awareness, it enriches their lives enormously. Many of these patients say that they perceive their cancer as a gift, and it is a gift they intend to appreciate for as long as possible. Despite the difficulties and horrors of surgery, radiation, and chemotherapy, many say it was all worth it. And they mean it. Amazing! This is not a platitude, but a spiritual awareness that they feel had to be forced upon them because they were too busy to achieve it in any other manner.

Similarly, the patient who has recovered from fibromyalgia or chronic fatigue syndrome sees life with different eyes. Prior to their illness, the majority of my patients with these symptoms were go-getters. They were highly productive, creative men and women (and boys and girls as well) who pushed the envelope a bit too far. When they heal, they rarely go back to their old habits—they have come to realize how precious life is and experience it differently now. They reprioritize their time: relationships may come first, and, though still creative and productive, they are not as driven as they once were (which they experience as a relief); instead, they now enjoy every aspect of their lives. This sense of balance is far more precious than they had realized. Their illness, now a thing of the past, provoked a change in what gives their lives meaning and purpose.

For want of a better term, I call this a Spiritual Awakening. I don't mean it as a religious event (although it can be), but as a way of seeing life in a completely new way.

We all are raised in a particular way, with parents, grandparents, teachers, religious instructors, and friends all (with good intentions) sharing their views of the world with us. Rarely does anyone take the time to help us understand our own unique viewpoints. Rather, we tend to look upon children as blank slates and think that we can help them see the world in the correct way—*our* way. With few exceptions, children accept this guidance unreservedly and have no reason to question whether this is the proper road for them. Why would they? And if nothing major comes along to make them question their values or purpose, they may go a whole lifetime on autopilot, guided by their mentors.

Please understand, I am not saying that everything our parents and others teach us is wrong—I know they are doing their best to raise us properly. What I am saying is that each of us is born a unique being, with unique gifts and talents, and we may not be aware that what we are doing is out of sync with our natural tendencies.

A glaring example would be the child of a prominent attorney who grows up expecting to take over her father's firm after she graduates from law school. This is a good thing, yes? But perhaps this child is more artistically or scientifically inclined, and only later in life, while working long hours at the law firm, does she wonder, "What am I doing here? I don't really like this career very much."

There are hundreds of examples of this sort of epiphany, but my point is simple: Many millions of human beings are toiling away at lives that are not very meaningful for them. Sometimes it takes a major illness to shake up the system before they finally ask themselves these vital questions, and this questioning opens up the opportunity for a meaningful life lived with purpose—from the perspective of that individual and that individual alone. This means that illness often does have meaning for those who are burdened with it, and if they can find that meaning, healing will be faster and more complete.

Of particular interest to me is to observe how differently illness is perceived by those who have it. I have several patients who are well known for their research, books, and lectures and who have been struggling with Lyme disease and/or mold toxicity for many years. Despite their illness, they plow forward, continuing to be of service to others. Other patients with similar symptoms and diagnoses are mired in illness and become bedridden. What is the difference between these two groups? I do not know with certainty. Possibly, the more functional group has a stronger constitution and can cope with illness better. However, having studied this population for many years, I am more inclined to believe that those who are more functional and productive have been able to push through their debilitating symptoms because their work and their lives are in better accord with their personal meaning and purpose.

I want to define spirituality not in religious terms, but simply as the energy that gives each individual meaning and purpose. I believe that this spiritual energy allows those fortunate beings who have found it and are working hard to live it (despite illness) to push through so that they are less affected by being unwell. Ultimately, it also makes them more likely to heal completely.

Accordingly, I have, for many years, asked my patients early on in treatment, "What gives your life meaning and purpose?" Those who have a ready answer are far more likely to respond well to treatment than those who look back at me with a blank expression and admit, "I just don't know."

Harnessing the Power of Affirmations

I have had many patients over the years who were aware of their psychological issues and underwent therapy, often for ten to twenty years or more. I have come to think that there is an erroneous fantasy underlying long-standing psychotherapy: that if you dig deep enough into the past, you will find one central trauma that created all the subsequent neuroses, and merely exposing this recovered trauma to the light of day will make it go away. While this may be true occasionally, I increasingly think that all that digging leaves you with little more than a lot of holes in the ground. I am not saying that psychotherapy does not have value, but I am of the opinion that unless it is undertaken with the *clear intention to change* whatever thoughts or behaviors are interfering with one's life, change simply will not occur.

Without ignoring how the past can shape the present, I have been impressed by the healing potential of affirmations. The basic concept is that the brain is simply a computer and does what the mind tells it to do. Unfortunately, we often have acquired "programs," if you will—things about ourselves that we have been told or taught at any early age, which we bought into and accepted, but may not be true. For example, if you drew a stick figure in first grade and your teacher or classmates rolled their eyes and told you that you had no artistic talent, you could go through the rest of your life believing that and never again try your hand at anything artistic. Some lucky individuals are later put in a position where they learn that their classmates were wrong and find deep fulfillment in artistic pursuits, while others remain isolated from that meaningful path. The issue here is that if you take in, without question, anything a parent, grandparent, teacher, priest, rabbi, minister, friend, relative, or acquaintance tells you, that statement can shape your view of yourself forever.

Far worse is that many of my patients have been told by their physicians, "You will have to learn to live with that. It can't be cured. There is nothing more we can do for you." If you take *that* in, it *will* limit your ability to heal. (Let me add that I hate that misuse of language in which an authority figure such as a physician says, "There's nothing more *we* can do." "There's nothing more *I* can do" is a fair statement. But the word *we* implies that no one on the face of this earth has anything else to offer—and surely that cannot be true.)

I have found affirmations to be a simple, inexpensive, and wonderful tool for overcoming such negativity and achieving spiritual rebooting. Basically, you write and say what you want to achieve, over and over again, and eventually your mind will reprogram your brain to make it happen.

There are a few rules about affirmations that need to be followed.

First, be as specific as possible. If an affirmation is at all vague, it lets the brain off the hook so that it doesn't have to follow the mind's command. So start each sentence with "I, _____ [fill in your name]," to be specific, and then state what you are commanding your brain to accomplish. In the example that follows, the words are carefully chosen. The word *relaxed* is clear. Using a term that is more vague, such as *happy* or *pain-free,* will not work. *Happy* is so nonspecific that the brain interprets it as, "Ah, you really don't know what you want yet, so

I will wait for a more specific command." *Pain-free* implies not having pain, and the word *not* (or anything negative) will backfire.

To explain: If I ask you *not* to picture a pink elephant, you will immediately picture a pink elephant. What's up with that? It turns out that in order not to picture a pink elephant, the brain must first picture the pink elephant and then cross it out. So any "command" phrased as a negative requires you to create an image of that statement before erasing it. *Pain-free,* therefore, requires you to visualize yourself in pain and then cancel that image, which leaves you right back where you started. It is in no way as positive as you might have thought.

Second, write each sentence in three different forms to cover the ways in which you may have received the original "program" or information. Writing, "I, _____ [your name]," includes only how *you* view this, and your difficulties were likely shaped by how *others* viewed it and you. Simple examples could be: "You are fat. You are stupid. You have no creativity. You are ugly. You are boring." If you internalized these statements, even if they were not true, you could be living them out on a daily basis. Therefore, write your affirmations in a positive voice (negative statements have no effect on the brain) in three different ways. I will give an example of what I often suggest to my patients; you can subtract anything that does not resonate with you or add additional words or phrases if you like:

I, _____ [your name], am relaxed, comfortable, and healthy.

You, _____ [your name], are relaxed, comfortable, and healthy.

He (She), _____ [your name], is relaxed, comfortable, and healthy.

You have to be consistent and do your affirmations every day—I recommend writing each affirmation, in all three forms, ten times daily. Having worked with affirmations for thirty years, I can assure you that your mind will do everything possible to test you to be sure you really mean it: "Oh, you poor dear. You've done so much today and worked so hard. Surely you can go to bed now and do this tomorrow?" If you fall into that trap, it won't work. Or your brain will argue with you: "I didn't realize you are a liar. You are not comfortable, you are not relaxed, and surely you are not healthy." You must answer with clarity: "Yes, I know it is not true now, but you will make it so. Just do it!" An affirmation is not a description of what is true now. It is a program—a command, if you will—to *make this happen.*

I encourage anyone who wants to try affirmations to follow this routine for at least ninety days. Your brain will not take this exercise lightly or accept it at face value unless you prove that you mean business. Your brain is merely an organ, and like all organs, it will do the least amount of work possible (not to be mean; that's how it is supposed to work), so for it to do a lot of work to change your thoughts and behaviors—it will have to shift your chemistry and structure, too, so this truly is a lot of work—you have to be resolute. If you aren't, your brain will not follow your program.

With persistence, however, I have seen affirmations do some amazing things in moving patients toward health.

Determining Your Own Sense of Meaning and Purpose

I cannot end this chapter without some final commentary on rebooting, if you will, spiritual awareness. Each individual must determine meaning and purpose in the context of his or her own life. My opinion of your choices, if I have one, does not matter. Rather, if anything brings you meaning and joy, it is worth pursuing. Even if you are wrestling with chronic fatigue and exhaustion, putting what limited energy you have into those pursuits usually pays dividends, actually giving you more energy over time and helping you get past the symptoms that are stymieing you.

If you truly cannot see anything in your current life that you are excited about, reflect upon your younger days. What did you enjoy doing? Did you go to church, play a musical instrument or sport, or have artistic interests or hobbies? Those would be great places to start your search. There must be more to your life than suffering and lying in bed waiting for healing to begin. You need to shift your focus to the things you can enjoy and appreciate.

I hope that I am placing sufficient emphasis on how important this process is for healing. Without it, healing often stalls out.

PART 4

CONCLUDING
THOUGHTS

My tale is mostly told. I would like to leave you with some thoughts about the kinds of attitudes that can promote or inhibit healing, which I hope will be helpful for physicians and patients alike, in Chapter 21. In Chapter 22, I let one of my patients tell her own story, which should be inspiring for those of you who have been ill for many years but keep believing that you will someday be healed.

This is followed by my Afterword, which goes even deeper into understanding why these epidemics are occurring now... and urges us all to bring this information into general consciousness so that we can do something about it.

CHAPTER 21

Is This a Cure for Everyone? Complicated Patients Versus Difficult Patients

Yes, Virginia, There Are People I Don't Know How to Help, or Can't Seem to Help, or Who Don't Seem to Want My Help, or (Gasp) Whom I Don't Want to Help!

I have spent years trying to treat difficult patients and making little to no progress. I have spent years trying to treat complicated patients and, after many years, have succeeded with most of them. I thought it might be useful to share what I have learned with my medical colleagues, and I hope this information will be helpful to patients as well. Some of this material describes difficult patients in an unflattering light, and although I do apologize for that, I hope it will spur some people to look within for the seeds of change.

After many years of accepting every complicated patient who came to my office seeking my help, I have evolved a screening process that allows me to talk with prospective patients and look for qualities I can work with.

First, I look for motivation: despite how long he or she has been ill, is this person still hopeful that he or she can get well and willing to do whatever it takes to achieve that goal? I know that, before ever consulting with me, many folks have spent a great deal of money on their health care and have little left to spend. Even so, those who will readily obtain all of the testing that I ask for (and I work hard to request only those tests that will give us immediately useful information) do much better than those who argue with me and debate the need for each and every test. Is it the patient who is motivated, or is it the patient's family and friends? Patients who arrive at my office solely because others thought I could help them do not do well. Patients who are referred by other physicians but have made no effort to read my books, blogs, or website or listen to my many podcasts generally do not do well. It's not that those sources of information are so wonderful, but that the patients are not trying to learn more about me so that they will know what they are getting into or will be able to relate to my information or the way I deliver it. Accepting someone else's referral without making any effort of their own is a passive activity that says a lot to me about those patients' motivation.

I also look for attitude and try to screen out patients who are needy or demanding (before they even step into my office). Patients who are unusually needy make it clear, with a barrage of emails that precede our first phone consultation, that they have such an intense need for care and attention that no mere mortal can fill it. When, in the past, I accepted those patients and began to work with them, they would send ten to fifteen emails a day coupled with innumerable phone calls to my office. Many of those calls and emails simply said, "Did you get my last email. I sent it ten minutes ago and you have not responded!" Patients who manifest these kinds of behaviors have unrealistic expectations of both me and my staff, and there is no way we can meet those expectations. Even if we try, most of these patients still get angry. My staff spends fruitless hours on the phone repeating the same information, which never seems to be enough.

Demanding patients, on the other hand, begin our conversations with unrealistic expectations that sound something like this: "You are one minute late for our phone consultation! Is this how you are going to treat me from now on?" My staff can never get their lab reports back to them quickly enough, or respond to their refill requests quickly enough, or schedule them the way they insist we must. I am old school: I am fiercely protective of my hardworking and compassionate staff. I am around them all day long, and I know how hard they try to meet everyone's needs with respect, as quickly as they can. If a patient cannot be grateful for their efforts, I cannot allow my staff to be abused verbally; I have a zero-tolerance policy for such behavior. I find that most of these patients feel self-righteously justified in their actions, and this just fuels their anger.

I also assess patients' understanding of their abilities. The majority of my patients have some degree of cognitive impairment. I know this. My staff knows this. We work tirelessly, repeating ourselves as often as necessary, to help them understand our instructions. A small set of patients with cognitive difficulties do not accept that they have cognitive difficulties and blame everyone else for what goes wrong. It is always my fault, or someone on my staff's fault, when they do not send in a specimen, or label it properly, or follow through with instructions, or misunderstand instructions, *which are handwritten at every visit*. They, too, have an anger that, though misplaced, is not to be tolerated. I am sorry that they are unable to think clearly, but until they own this liability, we can't work with them.

I look for "notchers" as well. In the Old West, gunslingers who survived a shootout would put a notch in their belt for every victory. "Notchers" is my name for patients who come to me having seen a wide variety of truly excellent practitioners, none of whom has helped one bit. In my younger days, when my ego was still up and running, I fantasized that I would help these people, demonstrating my superior clinical skills to all. With rare exceptions, was I ever wrong! After a difficult few months of effort, those patients left my practice, too, with one more notch in their belt. It took me a long time to realize that they really did not want my help; their goal was simply to prove that they, and their illness, were so special that no one could help them. A visit to me added weight to that self-fulfilling prophecy, and then they were off to see another wizard.

Finally, I try to screen out those patients who are chaos driven. Some people do not feel fully alive unless their world is falling apart. Patients with personality disorders, including narcissists, fall into this category. When you find such a patient, be aware that no matter how much progress he or she makes, he or she will likely sabotage it just when things look like they are turning around and, if you are not paying careful attention, will do so repeatedly. There will always be a new catastrophe that becomes a priority so that the patient cannot begin the treatment plan you outlined. After working with patients like this for many years, I finally understood the underlying dynamic and realized that I was part of the problem, not the solution.

I spent the better part of my forty-seven-year practice career feeling like I had a long way to go spiritually before I realized that I could not love, or appreciate, every single being who came into my office. Somehow, I had gotten the idea that I should. It took me a long time to realize that not everyone wants to be loved or appreciated, and that in making the effort, I was wasting their time and mine. One important spiritual quality is discernment, and I hadn't been using mine all that well.

I wrote this short chapter to communicate to my fellow healthcare practitioners that we need to turn a discerning eye to every interaction that seems odd or unusual. A negative interaction may not be your fault, although most healthcare providers are, understandably, caretakers, and we do a great deal of soul searching to be sure that the problem in that interaction was not ours. I have come to realize that continuing to provide care for patients who, for whatever reason, cannot accept that care in the spirit with which it is provided is not good for the patient or the provider. It's a lot like dating: once you are aware that the relationship is not going anywhere, it is cleaner and healthier to end it soon. Prolonging the relationship is unfair to both of you. Move on to healthier relationships. If you remain in limbo, you are limiting yourself, and your friend, from finding something better.

Although some of what I have written here can be construed as negative, it is not intended to be. It is intended to be clear. It is important to start all healthcare relationships with a mutual understanding of respect, a mutual goal of true healing, and a mutual agreement to work hard and for a long time to achieve it. If both parties are not coming to this unwritten contract with the same intentions, it will not work for either party.

This is an aspect of healing that is not often addressed, and I have begun this dialogue with the hope that others will continue it. Medical practitioners tend to blame themselves when interactions go astray, and I want to bring these interactions into a clearer light both to reduce practitioner guilt and to improve our understanding of these interactions. To maximize the time we spend in healing efforts, we need to be aware when those efforts are going astray and take steps to fix the problem.

> **" It is important to start all healthcare relationships with a mutual understanding of respect, a mutual goal of true healing, and a mutual agreement to work hard and for a long time to achieve it. "**

CHAPTER 22

Ending on a Hopeful Note

While I had thousands of stories to choose from, I think Betty's is of particular interest because she went from housebound to getting her life back. Here she tells her story in her own words.

BETTY'S STORY

When I found Dr. Nathan, I was forty-eight years old and had been sick for over sixteen years. Before I became ill, I had been a researcher at a national lab, an adjunct professor, and a member of a local dance company. I had been diagnosed with Lyme disease, suspected I had been exposed to mold, and had an extreme level of light sensitivity triggered by the antibiotic treatment for my Lyme disease. The light sensitivity manifested as a deep, sunburn-like burning on my face and the backs of my hands. Initially I was affected only by sunlight, but eventually regular room lights began to burn me. By the time I met Dr. Nathan, I had been living completely housebound and in the dark for seven years. I hadn't seen the sun in fourteen years. I used flashlights modified to be dimmer to navigate daily life, and all my windows were covered with blackout fabric that was taped at the edges. Even a few seconds' exposure to the light from a television or computer screen burned me. My husband had to remove the light bulbs from our refrigerator, dryer, and oven. My house was so dark, I couldn't see my hand in front of my face. It would take months to recover from an accidental flick of an untaped light switch. Just getting access to medical care was a challenge.

My problems started in 1996 with bilateral Achilles tendonitis that made it difficult to walk without pain. This was followed by chronic dizziness after a minor car accident in 1998, followed by bilateral carpal tunnel–like nerve pain that disabled me to the point where I was no longer able to work as an electrical engineer in 2001. In 2003, I tested positive for Lyme and the co-infections Ehrlichia, Anaplasma, Bartonella, and Babesia. After a three-day course of Zithromax for a diagnostic test, I developed a deep sunburn with minimal sun exposure. Three months after this initial reaction, the burning still had not resolved, but I began Lyme treatment and took antibiotics for four years while avoiding outdoor light. This resolved my carpal tunnel symptoms but did not resolve my chronic fatigue and joint pain. I never felt like I got well on the antibiotics, and I eventually discontinued them, hoping my light sensitivity would go away. It didn't.

I also tested highly positive for mold antibodies. In 2007, I moved and got rid of all of my belongings but did not experience any improvement. In 2010, I experienced an excruciating six-week period of burning on my face. In this time period, I also had to take a week of antibiotics for an infection, and that worsened my light sensitivity to the point where I had to avoid regular room lights and live in the dark, becoming completely housebound. In 2013, my vision began to oscillate. Any movement made the world look like it was bouncing up and down. In the spring of 2016, I began to experience crawling sensations on my skin. Daily life had been hard enough, but this symptom felt like my breaking point.

I tried supplements to help my methylation cycle, and I tried binders and glutathione in what I thought were small doses. I tried one sip of one drop of low-dose naltrexone in a glass of water. They all made me feel worse, and I had to discontinue them. Herbal treatments were tolerated better, but I was still sensitive and didn't recover. I was often puzzled that something would help initially but ultimately made me feel worse. Everything I tried seemed like a false start and a dead end. Nothing budged my light sensitivity. It felt like banging my head against a brick wall.

My symptoms were phototoxicity, oscillating vision, a feeling of eye pressure, chronic fatigue, body aches, joint stiffness in my knees, toes, and fingers, heart pounding, dizziness, light-headedness, history of fainting and presyncope, chemical sensitivity, post-nasal drip/chronic dry cough, intermittent low-grade fever, feelings of head/sinus pressure, low resistance to colds and flus, and crawling sensations. I also had low thyroid and adrenal hormones and high rheumatoid factor, and I was low in two IgG subclasses.

In late 2016, a friend told me about a book written by a woman who had also been living in the dark. I contacted the author and learned about mast cell activation syndrome. I started consulting with the author's nutritionist in England, who linked my light sensitivity with a histamine problem. I started on a low-histamine diet, histamine-reducing probiotics, and NeuroProtek LP. And they helped! I was able to get back on the computer, where I found Dr. Nathan. As soon as I watched an interview of Dr. Nathan's, I felt like I was one of his "sensitive patients." I emailed him, and his office's reply was so quick and so kind that it brought tears to my eyes. Most doctors weren't even willing to talk to me if I couldn't come into the office, but I could tell Dr. Nathan had read my medical history thoroughly, and he answered all of my questions. I was also very fortunate that my local doctor was extremely open to working with Dr. Nathan.

I started with Annie Hopper's Dynamic Neural Retraining System and found it tremendously beneficial. I immediately felt like my fatigue had been reduced by about 80 percent. I continued with the NeuroProtek, probiotics, and diet, and gradually added Claritin, ketotifen, Perimine, Allqlear, DAO, Chinese skullcap, Tox-Ease, Renelix, Itires, Viscum, and PSY-stabil. And—to my amazement—I started to recover. My light sensitivity improved, and the crawling sensations lessened. After my RealTime test came back positive for ochratoxins, trichothecenes, and gliotoxin, I started Saccharomyces boulardii, chlorella, and bentonite clay. For several months, I could tolerate only one drop of the bentonite clay, but from the first dose my oscillating vision started to go away. The idea that "less could be more" was a revelation to me. Eventually I was able to add cholestyramine, charcoal, and nystatin spray.

At first, I couldn't tolerate even the small doses that Dr. Nathan suggested. But slowly I was able to take more, and about six months into treatment, things really started to shift. I started with a string of red Christmas lights, then began to open my blacked-out window shutters a minute earlier every day. I became able to exercise. I was able to take some of the burden off of my husband, who was also severely ill with Lyme and mold illness. I cried from pure joy the first time I saw the sun again (while listening to The Cure).

Today, I have the lights back on in my house, and I am able to go for sunset walks in the park. I have been enjoying going out to dinner with my husband and reconnecting with old friends. I went to a movie and a play for the first time in seven years. I saw myself in the mirror for the first time in seven years. I can see my family again. I can see again. I have met my neighbors after living in my house for five years without ever having been seen. I figured they probably suspected my husband of making me up! I think I will never tire of looking at the sky.

There are, unfortunately, thousands of people like Betty whose lives have been disrupted completely by illness. Because her symptoms are unusual and her descriptions of them are vivid, few physicians were able to listen to Betty without dismissing her symptoms out-of-hand as being "obviously" psychological. The only thing obvious to me was that those physicians did not understand what was being communicated. Betty's debilitating sensitivity was not psychological in nature, but rather a consequence of mold exposure, Lyme disease, and its co-infections. Once understood, it could be treated properly. It is heartwarming for me to be able to help the Bettys of this world get their lives back.

I hope and pray that patients and healthcare providers alike will listen to this story and others like it and appreciate that no matter how disabled patients become, there is real hope that they can recover fully.

May this information resonate with you and enable you to find *your* path to complete healing.

Betty is happy and well today.

Afterword
A Plea: May We Come to Our Senses

I wanted to end this book on a hopeful note, and I have. But as I reflect on the message I am trying to communicate, I realize that having hope is not enough. We need to have wisdom as well.

As valuable a model for healing as this book may be, and as a paradigm that has the potential to ease the suffering of millions of people—a paradigm that can be expanded and evolved—it is only a stopgap. As I have reiterated throughout this book, we must always be looking for the deepest causes of illness so that true healing can occur. A good Band-Aid is still only a Band-Aid. It may be helpful for healing a wound, but shouldn't we address what is causing that wound, too?

Even if you were to take the information herein and utilize it to remove mold and other toxins and *Bartonella* and other infections and reboot all the systems of the body that are weak or failing so that all patients could recover their health and get their lives back, I still would not think I had done my job. I have not yet made a dent in the consciousness of the world I live in to help people understand that the causes of these conditions have yet to be addressed. The epidemics (barely acknowledged by conventional medicine) of chronic fatigue syndrome, fibromyalgia, autism, dementia, autoimmune illness, cancer, mold toxicity, and Lyme disease all share a common theme: inflammation created by an immune system that has been overstimulated and does not know how to correct itself.

Given the profound shifts in our gut microbiome created by an obsession with being germ-free, the overuse of antibiotics, and the insidious accumulation of toxins in our environment, we are all at risk of becoming ill. What we are seeing is the tip of the proverbial iceberg. Unless this problem is addressed—immediately—we are all in trouble. All of us. No exceptions.

I do not know how much time we have to get on top of this problem, but I fear we do not have the luxury of debating the finer points ad nauseam. We must set aside our personal agendas and come together to admit that we all have contributed to the creation of this toxic planet so that we can fix it. Addressing the excessive use of chemicals (most of which have never been tested for their effects on living beings), the accumulation of radioactive materials and heavy metals, the overuse of electromagnetic fields (which are not completely compatible with biological systems and energies), and global warming (a result of all these factors) must be a priority. Merely paying them lip service will not suffice.

Mother Earth is looking more and more like my chronically ill and sensitive patients. If I were to apply the paradigm expressed in this book to her, I would immediately find the specific chemical toxins that were making her sick, remove her from those exposures (how, I do not know), and detoxify her. I would find ways to reboot those systems that are not functioning properly so that she could thrive again.

Using different imagery from the perspective of Polyvagal Theory, we would realize that Mother Earth does not feel "safe" now. She is toxic, and we need to help her quiet down her vagus nerves by removing toxins and rebooting her nervous system.

We must come together to do this. We must reach out, talk with one another, and reestablish real connections to begin this process of neurological rebooting on a planetary level. Our infatuation with technology and our obsessive use of electronic devices has diminished this form of connection, not enhanced it. Our "friends" on Facebook are only virtual friends. While virtual reality is amazing, it is still virtual. A casual glance around us reveals thousands of people holding tiny boxes to which they are fixated, alone and isolated in a sea of electromagnetic stimulation. We are still human, and we require human touch and human connection to receive the neurological stimuli that can heal us. Hence, we must come back to our senses: the direct perception with all our senses of the natural world. Anything less is an illusion.

I pray that all of you will help in this healing effort. Our lives, and the lives of our children and grandchildren, depend on it.

Acknowledgments

First, to my incredible wife, Cheryl, whose constant encouragement, support, and unconditional love provide the foundation that makes everything possible.

To my three children, Aviva, Jules, and David, and three grandchildren, Avi, Anjali, and WiLo, all of whom are creating their own unique paths. I have often remarked that parenting is the most humbling experience available to us, a constant source of revelation about how little we know.

To our dogs present and past, Joey, Precious, Kai, and Jessie, who not only taught but embodied unconditional love (full disclosure: some of that love may have been a bit treat-driven), and our cat, Lucia, who demonstrated how long and comfortable life could be even with multiple disabilities.

To the contributors to this book, Robert Naviaux, MD; Joel Friedman, MD; John Banta, CIH; Kristine Gedroic, MD; Dave Ou, MD; Lauren Gordon-Fahn, MAOM, LAc; Jeff Greenfield, DO; and Bob Miller, MD; for their wisdom and experience and their willingness to share it in this book.

To my teachers, mentors, and friends who have contributed immeasurably to my knowledge and skill: starting with Elizabeth Kübler-Ross, MD, who taught me in medical school not only how to talk with patients, but more importantly how to listen to them. To Phil Curcuruto, DC, and Stan Weisenberg, DC, who taught me both intellectually and personally how emotions are stored in the body and how to release them. To Joe Barber, PhD, who taught me the critical importance of the precision of language in motivating patients to move toward healing. To GMA: Eric Gordon, MD; Wayne Anderson, ND; Alice Prescott-Sullivan, DO; Jeff Greenfield, DO; Annemieke de Lange-Austin, MD; Azra Mael, MD; and Alan McDaniel, MD; for scintillating and fabulous discussions as we searched for the answers to difficult questions. To Robert Gitlin, DO; Karla Gitlin, DO; Terry Turner, DO; and my wonderful staff at Redwood Valley Clinic,

Kasha, Debbie, Yiping, Michele, and Ema, without whom I could not continue to provide the kind of care our patients need. To Joe Brewer, MD; Patricia Kane, PhD; Ritchie Shoemaker, MD; Jacob Teitelbaum, MD; the ISEAI Board; Lawrence Afrin, MD; Steven Rochlitz, PhD; Joe Burrascano, MD; Rich Van Konynenburg, PhD; Tapan Adhya, PhD; Richard Deeth, PhD; Amy Yasko, PhD; Carolyn McMakin, DC; Len Ochs, PhD; and Stephen Buhner, who are my friends and teachers and mentors. I am blessed beyond measure for everything you have given me and for the opportunity to have you on speed-dial.

Without the enthusiastic and fabulous efforts of the Victory Belt staff, this book could not exist. Special thanks to the publisher, Erich Krauss, for his vision and commitment to bringing forth this information to reach the largest audience possible, and to my editor, Pam Mourouzis, for her brilliance in sharing my voice and helping to fine-tune this labor of love.

Finally, to all my patients and the thousands of wonderful folks who have touched my life and given me insights, direction, and love not specifically noted here (you know who you are).

As you can see, it takes more than a village to produce a book like this. It takes a whole universe.

Thank you.

APPENDIX A

Addressing Mold Problems
By John C. Banta

This appendix is for people who are really, really sensitive to mold and the people who make up their support teams. Its purpose is to give you a basic understanding of what causes mold problems to develop in homes and how to prevent them. This information should help you determine if a mold problem should be suspected, how to investigate it, and how to have mold remediation effectively performed, if necessary. Finally, it provides a plan for keeping mold issues under control.

As a Certified Industrial Hygienist who specializes in mold and water damage, after more than thirty years of investigating and making recommendations for fixing mold problems in homes and other buildings, I have come to a number of conclusions:

- Moldy homes can be fixed.
- Not everyone has the time, energy, budget, or expertise to fix a moldy home.
- The level of care and expertise necessary to address mold problems differs depending on the needs of the inhabitants.
- It is necessary for a home to work for the most sensitive person in the household.

A single appendix is not enough to answer every question or solve every mold problem, but I intend for it to provide a good start to help you control exposure to problematic types of mold. Having a basic understanding of the nature of mold can help you make the best decisions for dealing with mold growth.

Mold Basics

The principles for preventing mold from developing and for getting it back under control after it has grown aren't complicated, but they do require a dedicated approach.

Keep It Dry. The limiting factor for preventing mold growth is the amount of moisture present. Damp buildings develop mold growth. Dry buildings never grow mold. There will always be mold spores and fragments that migrate into a building from the outside, but unless there are elevated levels of moisture, they will not germinate and grow.

Keep It Simple. Once mold has begun to grow in a building, it is easy to become overwhelmed by the bewildering array of chemical formulas, products, methods, and devices that are touted as "the" solution to mold problems. A few are helpful. Some provide limited or minor benefits that are often way out of line with their price. Many can be downright dangerous or create more problems than they solve.

The basic principles of successful mold remediation are simple and straightforward:

- **Protect the people and pets living in or remediating the building.** They should be safeguarded throughout the process of returning the home to a safe and healthful environment.

- **Make a road map for getting from where you are to where you need to be.** The condition of the building should be documented before, during, and after mold remediation is performed. This documentation will provide a plan for addressing mold problems based on the individual levels of sensitivity of each inhabitant. It also will help determine what needs to be done if the remediation process is sidetracked by complications.

- **Isolate and control the contaminants whenever possible.** The mold should be isolated as close as possible to the area where growth has occurred to limit the spread of contamination. This is usually accomplished with the use of containment and airflow controls. The more widespread the contamination has become, the more work will be necessary to regain control.

- **Physically remove the mold.**

  - Water-damaged and moldy materials such as insulation and gypsum wallboard need to be removed and discarded.

  - Mold growth needs to be physically removed from wood framing surfaces via abrasive cleaning.

- Areas of the home and personal possessions that have been affected by settled spores and mold fragments should be thoroughly cleaned to remove those contaminants. It is necessary to clean to a level that is acceptable for the most sensitive person in the household, and to keep cleaning to that level to collect the spores and fragments that will continue to migrate into the living space.

  1. Hard-surfaced items are easily cleaned.

  2. Clothing and linens should be laundered with detergent or dry-cleaned, if necessary.

  3. Thick, soft, porous items like pillows, upholstered furnishings, and mattresses need to be evaluated on a case-by-case basis. It might be easier and more cost-effective to replace these items than to clean them.

- Attempts to kill, suppress, or encase mold are rarely successful and frequently make the condition worse.

- **Correct the moisture problem.** It is essential to correct or control the source of the moisture that caused the mold growth. If the moisture remains, the mold growth will return.

Understanding how mold grows and behaves in nature aids in understanding how a home can become infested with mold and what needs to be done to prevent and correct mold problems.

Mold in Nature

Various types of mold are commonly lumped together and considered one type of organism, but calling it all mold oversimplifies what is taking place. There are thousands of types of mold, and each type has a different nature. Some like warm conditions, and others grow at temperatures all the way down to freezing. There are types of mold that grow when humidity creates dampness, while other types require a wetting event in which the spores must soak for many days before they can germinate and begin to grow.

Once mold spores have germinated on a surface, the continued moisture level and the nutrients present determine whether the mold continues to grow or dies. Some types of mold need the nutrients present in green lumber that has recently been cut but has not yet dried and cured. Others require that the materials be predigested by other microorganisms. When one microorganism has used up the nutrients on which it thrives, other microorganisms take its place. Mold growth is usually an important part of this cycle. What occurs in nature is an incredibly complex and balanced ecosystem, with each type of organism occupying a niche but being kept under control by the environment and other nearby organisms. Some organisms cooperate, while others compete. It is truly a microscopic jungle out there, with a normal outdoor fungal ecology being the healthy status or condition that maintains a balance on the landmasses of our planet.

The types and quantities of mold that grow outdoors are frequently influenced by daily and seasonal outdoor temperatures. Most molds that grow outdoors year-round prefer cooler temperatures. They can survive in warmer conditions, but many can thrive only when it is cooler. In much of North America, nightly summer temperatures commonly drop into the low- to mid-sixties. This climate favors the outdoor types of mold commonly known as *leaf molds.*

The term *mildew* is often incorrectly used to describe mold problems in homes. Mildews are parasitic types of mold that grow on living plants. After the leaves die and fall from the plants, the mildews no longer have the necessary conditions to grow. Instead, common leaf molds begin to digest the nutrients they need. This is often the first step in turning leaves and other organic matter into soil. Many of these leaf molds are hardy, cold-loving organisms that are able to grow under conditions that do not favor the types of molds that grow in warmer conditions. Leaf molds begin breaking down leaves during autumn and into the winter months. They dominate the outdoor microflora throughout the year. Because the outdoor temperature tends to fall below room temperature almost every night, even in summer, these cold-loving outdoor molds remain the most common types of outdoor mold in most of North America.

The combination of temperature controls and competition between types prevents any particular type of outdoor mold from dominating. This tends to keep molds at normal levels. In cases where levels climb beyond what is comfortable, most people are able to keep seasonal allergy symptoms under control with over-the-counter allergy medicines. A growing number of people require rescue inhalers or other medications to help control asthma attacks, which can be predisposed by mold as well as other allergens.

Soil molds are the fungi that take over for leaf molds by completing the composting process and producing mulch. These molds grow at the warmer temperatures typically encountered as a compost pile heats up. The surface of the pile tends to remain cooler and be colonized by leaf molds. Soil molds favor the warmer temperatures located toward the warm, moist center.

Nature is patient. It takes over a year for a naturally occurring compost pile to form soil. As the leaves rot, the humus that is produced cools, and earthworms, bugs, and other insects help turn the compost from the bottom by pulling the newly produced soil downward into the earth. This means that the *Penicillium* and *Aspergillus* molds that are typically buried deep in the warm center do not become airborne to the same extent as the leaf molds found on the surface, where the process of composting is just beginning. Most of the spores floating around in the outdoor air are from leaf molds. Soil-producing mold spores also can be found in the outdoor air, but typically only as a minor component.

When people make compost, they typically turn the pile to mix the ingredients thoroughly and produce more uniform compost faster. Turning a compost pile results in a massive release of the same types of mold spores and fragments that are caused by water damage in homes and make really, really sensitive people sick.

The Mold Life Cycle

When a mold spore lands on a moist, nutritive surface, it begins to take on moisture. If that spore remains sufficiently damp for a long enough period, it will germinate, sending out a single thread-like structure called a *hypha*. If conditions are favorable, additional hyphae will form a matlike group of filaments called a *mycelium*. The presence of moisture allows additional acids and enzymes to be released to continue the process of turning digestible surfaces into absorbable nutrients for the mold.

If the environment is too dry, the spore, much like a plant seed, will be unable to germinate. This is because the outside of the spore is hard and durable, with the ability to repel water. In order to germinate, the spore has to soak long enough for the moisture to penetrate that coating. If the spore is wetted briefly, it may start to absorb water, but if it dries out again, it will remain dormant.

Once a spore has germinated, the mold must develop all the way to maturity and release new spores in order to reproduce and survive. If the spore geminates but dries out before new spores are formed, the spore will die. Adding water to an area of dead mold growth will not cause the dead mold to come back to life, but spores that have not yet germinated are able to germinate and grow in suitable conditions. A surface that already has mold growth on it also can develop additional growth faster because that surface is already partially digested. If the spore germinates and digests enough material to sustain its growth until it reaches maturity, then new spores are produced in the millions per square inch.

When a moldy surface dries, the mold does not die immediately; instead, it goes dormant. The dry mold is typically made up of a mixture of threadlike growth structures, which die off quickly, and more durable spores, which are seedlike structures that survive in a dormant state when conditions aren't right for growth. The spores can travel to new areas in a variety of ways. One of the most common is to be blown or launched into the air, where they float around. In still air, the largest spores will settle onto a nearby surface within a few minutes. The smallest spores may remain airborne overnight.

After the spores land on a surface, some types can remain dormant for years, waiting for moisture. Once the surface becomes damp or wet for a long enough time, the spore germinates. At first, the amount of available nutrients may be minuscule, but as the mold grows and spreads out into a matlike area, more nutrients are released as the surface is digested.

If the mold dries out and begins to go dormant again, some types begin to produce toxins called *mycotoxins* that can inhibit the growth of other types of mold and other microorganisms. Toxin production allows certain types of mold to stake out their territory and defend it from other types of mold attempting to invade. If a mycotoxin-producing mold has ideal growing conditions, little or no mycotoxin is needed or produced. When that mold is stressed by drying, it can turn on its mycotoxin production as it dries. In addition, if an invading mold tries to grow into the territory already occupied by a mycotoxin-producing mold, mycotoxin production can be stimulated. This is chemical warfare on a microscopic level.

How Do You Know if You Have a Mold Problem?

Mold isn't out to get us, but sometimes people are harmed. A severe mold problem can affect healthy people. Lesser amounts of mold growth can affect highly sensitive people. The types and quantities of mold and the individual sensitivities of the occupants influence the amount of mold that can be present in a home or office building without the occupants being adversely affected.

Just because 25 percent of the population has the genetic potential for mold sensitivity does not mean that they are demonstrating that sensitivity. The types and quantities of mold present may not have reached the concentrations needed to activate or overcome the trigger point necessary for the genetic potential to be activated and for mold to cause symptoms. An important goal is to keep mold exposures in homes at levels low enough not to trigger the on switch for those genes. If the sensitivity has been triggered, the mold levels need to be reduced to the point where they do not trigger adverse reactions.

As stated earlier, water or moisture is the key ingredient in mold problems in buildings. A home can get wet in many ways. Different types of wetting tend to create different kinds of mold conditions.

Prior to the energy crisis of the 1970s, buildings were rarely insulated and leaked a lot of air. They were much draftier than today's energy-efficient structures, and when they got wet, they tended to dry by themselves relatively quickly. Newer buildings do not have the same drying capacity. When an insulated, energy-efficient building gets wet, professional emergency measures usually need to be taken to dry it quickly. Modern building practices create inaccessible pockets of moisture that can remain long enough for mold spores to germinate and colonize the nutrients that are present in the construction materials. Hybrid buildings that are partially remodeled without appropriately upgrading moisture control measures are more likely to develop mold problems.

Each person exhales about a quart of water a day. When we add in the moisture emitted by cooking, cleaning, and bathing, it totals about a gallon of water per person released into our homes in the form of water vapor each day. If the home is unable to dry out the amount of moisture that enters, over time that moisture can build up to the point where mold growth is supported.

Buildings that have high levels of humidity favor condensation molds on cold surfaces and warm-loving mold growth on dusty surfaces. These molds typically do not require liquid moisture to germinate and begin growing. A few of these types of mold can sustain their growth at humidities over 60 percent and thrive at humidities between 70 and 80 percent.

When a home gets wet, it becomes a giant series of adjacent monocultures. Chronically wet gypsum wallboard sheathed in paper is the most common monoculture source of problematic mold growth in modern homes. When it stays wet for more than a few days, the paper linings on walls and ceilings favor the growth of massive amounts of mold, with millions of spores concentrated in every square inch.

We tend to keep our homes warm for comfort. These warmer temperatures favor the growth of soil molds in wet environments. A few spores of warm-loving mold will always be present in the settled dust on every square inch of surface in our homes. When these surfaces get wet and stay wet, those spores absorb that moisture and, within a couple of days, start to germinate and grow. Warmer temperatures support the types of mold that cause the most health problems. These are the types of molds that compete by either growing quickly or by producing toxic chemicals.

Cold-loving molds also can grow on cold surfaces when they are wet for multiple consecutive days. Condensation on cold windows or on the cooling coils in air conditioners and refrigerators can support the growth of cold-loving molds. For the most part, cold-loving molds are not the types that produce mycotoxins. The ability to grow at colder temperatures is enough of an advantage that they do not need to produce toxins in order to stake out their territory.

Dampness in buildings can result in many problems beyond mold. Dust mites are more common in damp homes, maybe because dust mites commonly consume mold. Mold and dust mites both need moisture and tend to grow under the same conditions.

Visible Mold Growth

Visible mold growth indicates that the indoor moisture level in the building is elevated. Visible growth should always be addressed; the source of the excess moisture must be corrected. If you see something that you think is mold, it is easy to collect a surface sample (tape lift) from the area of suspected growth and have it tested.

For mold to be visible to the naked eye, the growth has to be sufficiently dense; a density of 1 to 10 million colony-forming units per square inch is typically required. The standard of care for mold problems is that the building be dry with no *visible* mold. This is adequate for most people but may be insufficient for people who are hypersensitive to certain types of mold.

If the mold is from condensation moisture, it is likely a cold-loving leaf mold. This type of mold is usually observed on windows in winter and on cold-air supply register grilles in summer when the air conditioner is operating. An uninsulated cold water pipe also can form condensation moisture. If the water drips onto a nutritive material that is cold, then cold-loving leaf molds are more likely to develop. If the water drips onto room-temperature or warmer material, then the types of mold associated with water damage are more likely to develop.

Musty Odors and Hidden Mold

Musty odors are a strong indicator that moisture levels are elevated and mold is actively growing, even if you do not see any areas of growth. Mold produces musty odors only when it is damp and growing. Dormant or dead mold does not produce these gases. Once the area has dried, the odor disappears. Over time, the dormant mold will die off, but this does not mean the mold is gone or is no longer an issue. The EPA tells us, "Dead mold is still allergenic, and some dead molds are potentially toxic." If an area has musty odors but the source of the moisture has been corrected, it may be worth having the area investigated further to determine if hidden mold contamination is still present. It is becoming more widely recognized that some people have adverse reactions to the odors produced by actively growing mold.

A common misconception is that mycotoxins cause these musty odors. Actually, gases called microbial volatile organic compounds (mVOCs) cause the odors. Mycotoxins are not mVOCs, and they are typically odorless. Mycotoxins are attached to particles and filtered out by particle filters, whereas mVOCs are gases that pass right through particle filters. A quality HEPA vacuum will collect mycotoxins along with dirt and dust but is unable to remove musty odors.

Controlling mVOCs requires homes to be clean and dry. Using shortcuts like air purification, ozone, or other chemical treatments may reduce the odors, but it can lead to a false sense of security. The odorless particulates and the sources of mold growth still need to be properly remediated.

Mold growth is more likely to reach high densities where it is hidden from view because locations like wall cavities tend to dry more slowly than the exposed sides of walls. A history of water damage or previous mold growth frequently provides clues as to where mold may be hiding. It is common for mold growth to be camouflaged with paint or other cosmetic repairs. The growth remains dormant until the surface gets wet again, but these hidden sources can continue to release microscopic mold spores and fragments into the occupied space, and sensitive people can be affected. If the source of long-standing moisture is repaired but the mold is addressed only cosmetically, then it is likely that mold is still present.

Addressing Cold-Loving Molds

- Condensation on windows resulting in mold around the windows and sills. *Solution:* Dehumidification, ventilation. Clean up moisture and allow the surfaces to dry every day. Reduce the amounts of moisture coming into the home.

- Condensation on cold uninsulated exterior walls. *Solution:* Keep furniture and personal belongings several inches away from exterior walls to allow for air circulation and keep items off the floor so air can circulate underneath. Use fans, because air movement promotes evaporation. Use ventilation or dehumidification to promote drying of evaporated moisture. Reduce levels of moisture coming into the home.

- Condensation on exterior walls in closets. *Solution:* Don't over-pack closets; allow for air circulation. Never put anything damp (such as jackets or shoes) in a closet. Keep the closet door open or use a louvered-type door.

Addressing Warm-Loving Molds

- Water-damage molds grow in dust and on warm, damp surfaces. These types do not typically produce mycotoxins but can cause allergic reactions in sensitive individuals and exacerbate asthma. *Solution:* Warm air holds more moisture than cold air, so mechanical air conditioning can help remove excess humidity while it cools the home. Make sure the water is able to drain from the AC unit to a drain or the outside. Reduce the amount of water vapor that comes in and remove the excess by dehumidification.

- Short-term internal leaks: A sudden bursting of a pipe, fixture, or appliance is usually observed right away and can release a lot of water. *Solution:* Wipe up excess water immediately, but don't assume that you will be able to wipe up all the water that has flowed into nooks and crannies and inaccessible spaces. Professional drying involves the use of moisture meters, industrial fans, and dehumidification. The first twenty-four hours often makes the difference between an acceptable emergency response and a low-grade mold problem that can easily develop into a situation every bit as difficult to address as long-term water damage resulting in extensive mold growth. If you wait forty-eight to seventy-two hours to start drying things out, you are unlikely to prevent mold growth. Once a musty odor has developed, it is likely that mold remediation will be needed.

- Long-term internal leaks: A fixture or appliance that develops a slow drip or leak often goes undetected until a musty odor is detected, indicating mold growth. Slow leaks can allow water to migrate great distances. A slow drip into a wall cavity or concealed space can remain for weeks or months and result in massive amounts of damage. *Solution:* If an odor develops, don't use room air fresheners or sprays. Look for the source and have the issues addressed by a professional mold remediation company.

- Homes with external leaks are often similar in scope to buildings with long-term internal leaks. Each type of construction has its own potential problems that could fill a book, but here are some of the more common issues:

  - Roofs: Have routine inspections and keep damage from developing. If a leak is discovered, have it promptly addressed and treat it like an emergency. By the time water has soaked through insulation and ceilings to a point where it can be observed, there is a good chance that mold has already developed.

  - Exterior walls: Have routine professional inspections to catch problems early. Never let mulch, soil, old leaves, plants, or any other debris build up against walls. There should be a 6-inch space between the bottom edge of the stucco or siding and the top of the soil or mulch. If an adequate distance is not maintained, water damage will result. Have walls routinely inspected and repaired by an expert. Minor cracks may not be a problem, but major cracks that funnel water into the walls are. Brick walls that are properly installed and maintained can last significantly longer than siding or stucco, but incorrect installation can be a nightmare to fix and maintain.

  - Flashings must be properly installed to direct water away from internal components.

  - Gutters must direct water away from the building.

  - Foundations: Crawlspaces and slabs should be dry and properly maintained. Debris on the soil, earth/wood contact, elevated moisture, and other issues can lead to mold growth that can be brought by stack effect and other forces into the living space. Finished basements or subterranean walls with improper or failing moisture control can have lots of problems with hidden mold. Slab foundations frequently have plumbing installed in the slab. If it develops a leak, the water may migrate into adjoining walls, floors, and personal possessions.

Routine Cleaning

The history of the situation is important in determining whether mold should be cleaned up as part of routine maintenance and cleaning or whether professional assistance is required.

Minor mold growth such as occurs in windows from condensation or in tile grout requires routine cleaning, but not remediation. Regular cleaning can prevent buildup that is likely to cause adverse reactions in mold-sensitive people. Many sensitive people learn what level of proactive cleaning is necessary to prevent a mold problem from making their symptoms worse. Nontoxic tub, toilet, and tile cleaners work exceptionally well when used for routine cleaning of hard nonporous surfaces where condensation or surface moisture results in minor mold growth. Soap and water or detergent-based cleaners are best.

I often hear people say that they use only vinegar, baking soda, essential oils, or some expensive specialized cleaner designed to kill mold. This is simply unacceptable for addressing mold growth, spores, fragments, or mycotoxins. Trying to kill mold is not the answer and usually leads to more problems than it solves. Vinegar, baking soda, and essential oils have some beneficial uses, but ridding a home of mold is not one of them.

Always use a properly functioning HEPA-type vacuum. (Unfortunately, many HEPA vacuums are not functioning properly.) My favorite is the Shark Rotator Lift-Away with complete seal technology. At the time of this writing, I have tested more than 150 Sharks that were less than two years old and have found them to be consistently reliable. But even they have an issue that sensitive people have to overcome: they are bagless. It is important that the collection chamber be emptied outdoors, by someone who is not sensitive to the dust that will be released.

A HEPA vacuum can remove loose mold spores and dry fragments, dirt, and debris from surfaces. If you start with damp or wet cleaning, it makes mud, which is more difficult to clean. After the home and personal possessions have been vacuumed, hard surfaces that won't be damaged by moisture can be wiped with a disposable microfiber cleaning cloth that has been moistened (not soaked!) with a soapy water solution. The area must then be dried quickly and completely. If you use a minimal amount of moisture, the surfaces should dry by themselves in just a few minutes.

This approach to cleaning will remove surface mold growth, but not stains. Once the moisture problem has been resolved, a hydrogen peroxide–based cleaner can be effective for reducing or eliminating stains. It may be necessary to use a stain blocker before repainting walls that have had surface growth. These methods are effective only for mold that has developed on a surface film of condensation. They will not take care of what is hidden mold growth inside a wall cavity or behind tile and grout. If the water has gotten into a wall cavity or behind tile and grout, then merely removing the surface growth or stain will be insufficient. Removal of the moldy materials is necessary. If the source of moisture is not resolved, the mold will continue to reoccur.

Minor areas of growth may be able to be cleaned by a family member who is not hypersensitive. If mold growth is extensive, or if one or more people living in the home are hypersensitive, then professional remediation is of greater importance.

If you have adverse health effects that you attribute to mold, or your physician suspects mold exposure, seek professional assistance. Similarly, if you have a history of water damage that was not properly and quickly addressed, seek help from a professional.

Working with a Professional Mold Inspection or Remediation Company

There are do-it-yourself test kits and laboratories that work directly with the public, but an in-person inspection by a qualified mold inspector is usually an extremely important step. In my experience, without a physical inspection and testing designed to answer specific questions, a shotgun approach to testing is frequently misleading. The way the testing is conducted has a significant influence on the interpretation of the results. There are also many different types of testing, and each has its advantages and disadvantages.

Unfortunately, most mold inspectors have little or no training in working with people who have environmental sensitivities. Most mold remediation companies have even less training in this area. People suffering from environmental exposures often need a more experienced and knowledgeable mold inspector and remediation company.

Look for companies with at least five years of experience working with people with mold sensitivities. Those with experience only in insurance or real estate transactions are typically unable to provide the level of care needed. If experienced environmental consultants and remediators are not available in your area, it may be possible for an experienced consultant to work long-distance with your local resources.

A team approach is often best for a successful outcome. The other critical component is you. It is important not to turn over your life and the decision-making power to an expert. The expert should advise you so that you can make the best and most appropriate decisions for the health and well-being of yourself and your family.

The words *testing* and *sampling* are often used interchangeably, but there are important differences. Testing tells you what is present in the area in which the test is conducted. If you test an area of mold growth on a wall, you can find out, within the limits of the test, what type(s) of mold are present in that small area. The results do not tell you anything about what is 2 inches or 2 feet away. Sampling provides a way to predict the answers to a question based on a number of tests. The more tests you collect, the more you know about the situation, but of course the more expensive the determination.

A tape lift on a windowsill may reveal the presence of *Cladosporium,* a common outdoor mold found growing on condensation moisture in cold windows. Keeping it under control requires routine maintenance and cleaning and a reduction in the humidity levels in the home. That tape lift would miss the fact that water got into the wall around and below the window and resulted in mold growth that needs to be remediated by opening the wall cavity using appropriate controls to prevent the contaminants from spreading. In many cases, a visual inspection by a trained observer may reveal problems that need attention.

A mold inspector should take many steps before determining a sampling strategy. Many different types of samples can be collected, and each type has its advantages and disadvantages. The inspection and history of a building is often the most important part of addressing mold problems. Appropriate sampling can then be used to answer specific questions about the areas to be remediated and ensure the success of the remediation.

Again, this appendix is only intended to provide a brief introduction to investigations. For more information, see my books *Mold Controlled* and *Prescriptions for a Healthy House: A Practical Guide for Architects, Builders & Homeowners.* Additional information is available at www.JohnCBanta.com.

APPENDIX B

Supplement Reference Guide

This is not a comprehensive list of all the supplements that could be of value for the treatment of the illnesses discussed in this book, but rather a listing of some of the supplements that I have found to be of particular benefit for my patients. Other physicians will have their own favorites that may be equally helpful. You should trust the judgment of the healthcare providers who are guiding you.

I will present these materials by noting the company that makes them along with a brief listing of the products I use the most:

- **Beyond Balance** offers herbal extracts that are a bit unique in that they have a glycerine base, which makes them easier for many patients (especially children) to take than some of the tinctures that are alcohol based.

  One line of these products includes extracts that are specifically for some of the infectious agents this book has discussed: these include MC-BB-1 and BB-2 for Lyme borreliosis; MC BAR-1 and BAR-2 for *Bartonella*; MC BAB-1, BAB-2, and BAB-3 for *Babesia*; treatments for viral infection, including IMN-V, V-II, and V-III; MC-PZ and PARALLEVIARE for parasites; and detoxification materials, including TOX-EASE, TOX-EASE GL, and TOX-EASE BIND. I use a good deal of MC-BFM for dissolving biofilm, and I use MYCOREGEN for improving the immune system's response to fungal infection.

- **BioResource Inc.** distributes several homeopathic lines of products from Germany. The Pekana remedies include PSY-stabil to calm an anxious or overexcited nervous system, SomCupin for sleep, Itires to help detoxify the lymphatic system, Renelix to help detoxify the kidneys, and apo-Hepat for the liver (which is stronger). The sanPharma line includes Candida nasal spray (which I find helpful for sinus fungal infections) and MucorSAN and NotaSAN for inflammation, among many others.

- **Byron White Formulas,** along with the Beyond Balance formulas, have become the go-to tinctures to assist with the treatment of a wide variety of infections. A-Bab for *Babesia,* A-Bart for *Bartonella,* A-FNG for fungal infections, A-L Complex for Lyme, A-V for viral infections, and A-P for parasites are all of great benefit.

- **Deseret Biologicals** makes several homeopathic remedies specifically for some of the infections that affect toxic and sensitive patients; I have found their *Bartonella* and Lyme kits to be of particular value.

- **Klaire Labs** makes InterFase Plus, which, along with Beyond Balance's MC-BFM, is one of my favorite biofilm-dissolving agents.

- **Natural Immunogenics** makes Argentyn 23, a hydrosol silver material that I use extensively in both nasal spray and oral form for mold and Lyme patients to improve their response to antibiotics and antifungals. The "23" refers to the particle size of the silver, telling us that it is safe to use, even intravenously, without accumulating in body tissues.

- **NeuroResearch** provides high-quality amino acid supplements to help stimulate the body to make the correct balance of neurotransmitters such as serotonin, epinephrine, and norepinephrine. The combination of NeuroReplete and CysReplete has been of particular value for this purpose.

- **NutraMedix** distributes tinctures made by Lee Cowden, MD, many of which are specific for Lyme (Banderol and Samento) or *Bartonella* (Houttuynia and Cumanda) or are helpful for detoxification (Sealantro and Parsley).

- **Researched Nutritionals** makes many products; the ones I use most are Tri-Fortify liposomal glutathione (for performing the provocation test for urine mycotoxins), Transfer Factor Enviro (to improve the immune system's ability to respond to fungal infection), ATP Fuel for energy, and CytoQuel to reduce inflammation.

- **Xymogen** makes OptiFiber Lean, which I use as a gentle binder for mold toxins.

- **Zhang Chinese Herbs** include HH capsules *(Houttuynia)* for *Bartonella* and allicin, a potent form of garlic, among others.

Suggested Reading

Previous Books by Neil Nathan

Healing Is Possible: New Hope for Chronic Fatigue, Fibromyalgia, Persistent Pain, and Other Chronic Illnesses. Laguna Beach, California: Basic Health Publications, Inc., 2013.

Mold & Mycotoxins: Current Evaluation & Treatment 2016. E-book.

On Hope and Healing: For Those Who Have Fallen Through the Medical Cracks. Et Alia Press, 2007.

Chapter 3: Mold Toxicity

Baker-Laporte, P., E. Elliott, and J. Banta. *Prescriptions for a Healthy House: A Practical Guide for Architects, Builders & Homeowners,* 3rd Edition. Gabriola Island, British Columbia: New Society Publishers, 2008.

Brewer, J. H., D. Hooper, and S. Muralidhar. "Intranasal Antifungal Therapy in Patients with Chronic Illness Associated with Mold and Mycotoxins: An Observational Analysis." *Global Journal of Medical Research* 15, no. 1 (2015): 29–33.

Brewer, J. H., D. Hooper, and S. Muralidhar. "Intranasal Nystatin Therapy in Patients with Chronic Illness Associated with Mold and Mycotoxins." *Global Journal of Medical Research* 15, no. 5 (2015).

Brewer, J. H., J. D. Thrasher, and D. Hooper. "Chronic Illness Associated with Mold and Mycotoxins: Is Naso-Sinus Fungal Biofilm the Culprit?" *Toxins* 6, no. 1 (2014): 66–80.

Brewer, J. H., J. D. Thrasher, D. C. Straus, R. A. Madison, and D. Hooper. "Detection of Mycotoxins in Patients with Chronic Fatigue Syndrome." *Toxins* 5, no. 4 (2013): 605–17.

Crago, B. R., M. R. Gray, L. A. Nelson, M. Davis, L. Arnold, and J. D. Thrasher. "Psychological, Neuropsychological, and Electrocortical Effects of Mixed Mold Exposure." *Archives of Environmental Health* 58, no. 8 (2003): 452–63.

Delgado, M., and D. Ganea. "Vasoactive Intestinal Peptide: A Neuropeptide with Pleiotropic Immune Functions." *Amino Acids* 45, no. 1 (2013): 25–39.

Foster, J., P. Kane, and N. Speight. *The Patient's Detoxx Book.* BodyBio, 2002.

Gray, M. R., J. D. Thrasher, R. Crago, R. A. Madison, L. Arnold, A. W. Campbell, and A. Vojdani. "Mixed Mold Mycotoxicosis: Immunological Changes in Humans Following Exposure in Water-Damaged Buildings." *Archives of Environmental Health* 58, no. 7 (2003): 410–20.

Hope, J. "A Review of the Mechanism of Injury and Treatment Approaches for Illness Resulting from Exposure to Water-Damaged Buildings, Mold, and Mycotoxins." *The Scientific World Journal,* Vol. 2013.

Kahlon, T. S., et al. "*In Vitro* Binding of Bile Acids by Okra, Beets, Asparagus, Eggplant, Turnips, Green Beans, Carrots, and Cauliflower." *Food Chemistry* 103 (2007): 676–80.

Ponikau, J. U., D. A. Sherris, E. B. Kern, et al. "The Diagnosis and Incidence of Allergic Fungal Sinusitis." *Mayo Clinic Proceedings* 74, no. 9 (1999): 877–84.

Shoemaker, R. C. *Mold Warriors.* Baltimore, Maryland: Gateway Press, 2005.

Shoemaker, R. C. *Surviving Mold: Life in the Era of Dangerous Buildings.* Baltimore, Maryland: Otter Bay Books, 2010.

Chapter 4: Bartonella Infection

Buhner, S. H. *Healing Lyme Disease Coinfections: Complementary and Holistic Treatments for Bartonella and Mycoplasma*. Rochester, Vermont: Healing Arts Press, 2013.

Buhner, S. H. *Healing Lyme: Natural Healing of Lyme Borreliosis and the Coinfections Chlamydia and Spotted Fever Rickettsiosis*, 2nd Edition. New Mexico: Raven Press, 2015.

Horowitz, R. I. *Why Can't I Get Better: Solving the Mystery of Lyme & Chronic Disease*. New York: St. Martin's Press, 2013.

Strasheim, C. *New Paradigms in Lyme Disease Treatment*. South Lake Tahoe, California: BioMed Publishing Group, 2016.

Chapter 5: Mast Cell Activation Syndrome

Afrin, L. B. *Never Bet Against Occam: Mast Cell Activation Disease and the Modern Epidemics of Chronic Illness and Medical Complexity*. Bethesda, Maryland: Sisters Media, 2016.

Molderings, G. J., S. Brettner, J. Homann, and L. B. Afrin. "Mast Cell Activation Disease: A Precise Practical Guide for Diagnostic Workup and Therapeutic Options." *Journal of Hematology Oncology* 4 (2011): 10.

Chapter 6: Porphyria

McKeown, P. *The Oxygen Advantage: Simple, Scientifically Proven Breathing Techniques to Help You Become Healthier, Slimmer, Faster, and Fitter*. New York: HarperCollins, 2015.

Rochlitz, S. *Porphyria: The Ultimate Cause of Common, Chronic & Environmental Illnesses*. Cottonwood, Arizona: Human Ecology Balancing Science, 2013.

www.cpnhelp.org

Chapter 8: Rebooting as a Model for Treatment: The Cell Danger Response

Naviaux, R. K. "Antipurinergic Therapy for Autism—An In-Depth Review." *Mitochondrion* (2017). E-pub ahead of print.

Naviaux, R. K. "Metabolic Features of the Cell Danger Response." *Mitochondrion* 16 (2014): 7–17.

Naviaux, R. K., J. C. Naviaux, K. Li, A. T. Bright, W. A. Alaynick, L. Wang, A. Baxter, et al. "Metabolic Features of Chronic Fatigue Syndrome." *Proceedings of the National Academy of Sciences*, 113, no. 37 (2016): E5472–E5480.

Chapter 9: Rebooting the Nervous System

Bredesen, D. *The End of Alzheimer's: The First Programme to Prevent and Reverse the Cognitive Decline and Dementia*. United Kingdom: Penguin, 2017.

Brooks, R. E., ed. *Life in Motion: The Osteopathic Vision of Rollin E. Becker, DO*. Oregon: Stillness Press, 1997.

Doidge, N. *The Brain That Changes Itself: Stories of Personal Triumph from the Frontiers of Brain Science*. New York: Penguin Books, 2007.

Doidge, N. *The Brain's Way of Healing: Remarkable Discoveries and Recoveries from the Frontiers of Neuroplasticity*. New York: Penguin Books, 2016.

Hinz, M., A. Stein, and T. Uncini. "Relative Nutritional Deficiencies Associated with Centrally Acting Monoamines." *International Journal of General Medicine* 5 (2012): 413–430.

Hopper, A. *Wired for Healing: Remapping the Brain to Recover from Chronic and Mysterious Illnesses*. Victoria, British Columbia: The Dynamic Neural Retraining System, 2014.

Kidd, R. F. *Neural Therapy: Applied Neurophysiology and Other Topics*. Ontario, Canada: R. Kidd Books, 2005.

Larsen, S. *The Healing Power of Neurofeedback: The Revolutionary LENS Technique for Restoring Optimal Brain Function*. Rochester, Vermont: Healing Arts Press, 2006.

McMakin, C. R. *Frequency Specific Microcurrent in Pain Management*. Elsevier Ltd., 2011.

Meinig, G. *Root Canal Cover-Up*. Price Pottenger Nutrition, 2008.

Porges, S. W. *The Pocket Guide to the Polyvagal Theory: The Transformative Power of Feeling Safe.* New York: W. W. Norton & Co., 2017.

Rosenberg, S. *Accessing the Healing Power of the Vagus Nerve: Self-Help Exercises for Anxiety, Depression, Trauma, and Autism.* California: North Atlantic Books, 2017.

Schwartz, J., and B. Beyette. *Brain Lock: Free Yourself from Obsessive-Compulsive Behavior.* New York: ReganBooks, 1996.

Chapter 10: Rebooting the Immune System

Dosch, P., and M. Dosch. *Manual of Neural Therapy According to Huneke.* New York: Thieme New York, 2007.

Douglass, W. C. *Into the Light: Tomorrow's Medicine Today.* Panama City: Rhino Publishing, 2003. Georgia: Second Opinion Publishing, 1993.

Nambudripad, D. S. *Say Good-bye to Illness,* 3rd Edition. Buena Park, California: Delta Publishing Company, 2002.

Shallenberger, F. *Principles and Applications of Ozone Therapy—A Practical Guide for Physicians.* Self-published, 2011.

Williams, L. *Radical Medicine: Cutting Edge Natural Therapies That Treat the Root Causes of Disease.* Rochester, Vermont: Healing Arts Press, 2011.

Chapter 11: Rebooting the Endocrine System

Cohen, S. *Thyroid Healthy: Lose Weight, Look Beautiful and Live the Life You Imagine.* Dear Pharmacist, Inc., 2014.

Jefferies, W. McK. *Safe Uses of Cortisol.* Springfield, Illinois: Charles C. Thomas Publisher, Ltd., 1996.

Morgenthaler, J., and J. V. Wright. *Natural Hormone Replacement for Women Over 45.* California: Smart Publications, 1997.

Shippen, E., and W. Fryer. *The Testosterone Syndrome: The Critical Factor for Energy, Health and Sexuality—Reversing the Male Menopause.* New York: M. Evans and Company, 2001.

Wilson, E. D. *Wilson's Thyroid Syndrome: A Reversible Thyroid Problem,* 4th Edition. Lady Lake, Florida: WilsonsThyroidSyndrome.com, 2001.

Chapter 13: Reversing Weight Gain

Galland, L. *The Fat Resistance Diet: Unlock the Secret of the Hormone Leptin to: Eliminate Cravings, Supercharge Your Metabolism, Fight Inflammation, Lose Weight & Reprogram Your Body to Stay Thin.* New York: Broadway Books, 2005.

Gundry, S. R. *The Plant Paradox: Hidden Dangers in "Healthy" Foods That Cause Disease and Weight Gain.* New York: HarperCollins, 2017.

Mosely, M., and M. Spencer. *The FastDiet,* Revised & Updated Edition. New York: Simon & Schuster, 2014.

Chapter 16: How Your Genetic Makeup May Affect Your Ability to Detoxify

Mangan, P. D. *Dumping Iron: How to Ditch This Secret Killer and Reclaim Your Health.* Phalanx Press, 2016.

Chapter 17: Rebooting Multiple Chemical Sensitivities (MCS)

Ashford, N., and C. Miller. *Chemical Exposures: Low Levels and High Stakes.* New York: John Wiley & Sons, 1998.

Chapter 18: Rebooting How We Handle Stress

Peterson, J. *The Natural Bliss of Being.* Self-published, 2014.

Chapter 19: Rebooting the Psyche: Emotional Awakening

Walsh, W. *Nutrient Power: Heal Your Biochemistry and Heal Your Brain,* Revised and Updated Edition. New York: Skyhorse Publishing, 2014.

Index